**BLOOD, DRA**

# BLOOD, DRAGONS AND LIONS

How alienation and science led me to spiritual enlightenment and innovation

## KIN F. KAM

Inclusive Innovations

Copyright © Kin F. Kam, 2015

First published in 2015 by Inclusive Innovations Ltd

All rights reserved. No part of this book may be reproduced, adapted, stored in a retrieval system or transmitted by any means, electronic, mechanical, photocopying, or otherwise without the prior written permission of the author.

The rights of Kin F. Kam to be identified as the author of this work have been asserted in accordance with the Copyright, Designs and Patents Act 1988.

A CIP catalogue record for this book is available from the British Library.

ISBN 978-0-9933364-0-9

Cover design by Spiffing Covers Ltd

Prepared and printed by:

York Publishing Services Ltd
64 Hallfield Road
Layerthorpe
York YO31 7ZQ

Tel: 01904 431213

Website: www.yps-publishing.co.uk

*To all the defiant stoics, laughing and fighting lonesome battles in the face of adversity.*

# Contents

## Part I

1. **Invent, Therefore I Am** — 1
   *Austerity, freedom and risk* — 1
   *I want to be normal* — 5

2. **Islands and Seas** — 7
   *A Chinese immigrant in a foreign Chinese land* — 11
   *From SS Inchearn to Atreus* — 12
   *Memory of early childhood* — 16
   *What is H?* — 26
   *How do you get H?* — 28
   *Rebel without a cause* — 30
   *From Hong Kong to England* — 48

3. **Life in a Northern Town** — 55
   *Initial perception* — 57
   *Steep learning curve* — 62
   *Futile defiance* — 71
   *Learning to walk* — 79
   *Salt and vinegar?* — 87
   *A Chinese kid in an English school* — 89
   *Hope of the sojourned lights!* — 101
   *China! When will you wake up?* — 105

| | | |
|---|---|---|
| **4.** | **Science, Race and Culture** | 111 |
| | *The inspiration of Einstein* | 111 |
| | *Amazing calculus* | 113 |
| | *Race and culture* | 115 |
| | *Triumph and tragedy* | 124 |
| | *The British Chinese* | 126 |
| **5.** | **Freedom and Sufferance** | 129 |
| | *In the shadow of disasters* | 129 |
| | *Physics and independence* | 132 |
| | *God does not play dice* | 139 |
| | *China; hope and despair* | 143 |
| | *The Cambridge experience (and about Myjaugoek)* | 149 |
| **6.** | **What is Life?** | 161 |
| | *The first day* | 161 |
| | *Nuclear fusion* | 165 |
| | *The relentless vicious circle* | 169 |
| | *Natural mystic* | 171 |
| | *God and self realisation* | 177 |
| | *China's redemption* | 190 |
| **7.** | **Return to God's County** | 194 |
| | *Take it easy!* | 195 |
| | *Yorkshireman and the motorbike* | 199 |
| | *A dot-com adventure* | 206 |
| | *When demons fight* | 215 |
| | *Chinese or English? The identity question* | 221 |

## Part II

**8. An Inventor's Journey**    231

   *Three Inventions, one dream*    231
   *A year of searching and hope*    240
   *Hello Dragons!*    252
   *Paris, Smartcards and Hollywood*    267
   *Good news, bad news*    274
   *Annus Mirabilis?*    278
   *Chinese American vs. British Chinese*    291
   *Patent is not everything*    294
   *An interlude of normality*    296
   *Optimism and creativity*    301
   *Progress is about Inclusive Innovations*    305
   *Hong Kong, 10 years on*    312

**9. The Long Salvation**    317

   *Battles begin*    317
   *New blood, new knee, new life*    327
   *Dragons' Den Sport Relief*    338
   *Great expectation*    345
   *Innovation is the only way*    348

**10. Invent, Therefore We Are**    351

   Endnotes    361
   Acknowledgements    379
   About the Author    380

# Part I

# 1. INVENT, THEREFORE I AM

## Austerity, freedom and risk

*Mid-afternoon, 30 September 2004.* I press the "Enter" key on the PC's keyboard. A computer simulation of an electromagnetic wave starts to propagate from the left hand side of the monitor's screen, moving gracefully to an imaginary barrier. It was a simple model with many applications – one was to investigate how radio waves from mobile phones interacted with the human body, while another application was more sinister, to design a skin barrier for spy planes and unmanned drones which are invisible to radar waves. I didn't think much about how my research will be used; I just wished the damn program would work. The wave hit the simulated barrier, part of it was reflected and part of it transmitted; both waves then continued their relentless paths to the edges of the screen. The waves arrived at their opposite sides and I could see them gradually being absorbed by the wall; just as I wanted. YES, YESsss! I exclaimed very loudly, triumphantly punching my fists in the air and simultaneously jumping out of my chair with great joy. My painstakingly written program had finally worked perfectly and stopped crashing with those awfully familiar error messages. I had finally gotten rid of that bloody computer bug. Coincidentally, my boss Professor Marvin was just passing my office and he must have been perplexed by my joy since in only a few hours time my post-doc contract as a research physicist would come to an end and I would be unemployed.

In a way, I was rather relieved to be freed from the obligations of being an employee, as my work involved a lot of programming which I found tedious, and was frankly not particularly adept at. I love learning about nature, especially through the discipline of theoretical physics, but really hated coding computer programs. This is rather unfortunate for me as coding is pretty much a prerequisite for almost all physics related occupations. From now on, I am going to pursue what I want to do. The stage for my new career had in fact already been prepared; in less than three weeks time, I would enter the den.

*19th October 2004*: After waiting for over six hours in a cold draughty room, it was finally my turn to meet my potential business saviours. But within a few minutes, I was thinking to myself, what have I got myself into, standing in front of the stony faced investors of Dragons' Den, about to announce to the nation in a brand new reality business program: "Hi dragons, my name is Kin Fai Kam, I am a physicist and an inventor; my innovation is...." But, before I could finish my well-rehearsed investment pitch in front of the apparently rather egotistical, smirking millionaires, my speech became increasingly disjointed with breathlessness. I didn't understand why I was desperately grasping for air with every additional word now 'puffing' from my mouth. I had never felt like this before! Oh No, I thought, is this what a panic attack feels like?

In fact, my predicament was engineered by the BBC. Prior to the dubious privilege of appearing in front of the Dragons, one has to surmount a scarily steep and precarious staircase. I now realise this is a trick of the trade... to knacker the poor entrepreneur before being set on by the dragons! Well, at least all my anxieties of pitching to the dragons were now secondary to literally overcoming these stairs. In fact, they were potentially lethal as there was no hand rail for the last few steps; a situation which has since been remedied in subsequent series. A momentary thought entered my mind; if I accidentally fall down these stairs, I'd probably get the money I came for, except it will be from the BBC compensation funds instead of from the Dragons own pockets.

Hey, why make a mountain out of a few steps? Surely they're no big deal for an apparently able-bodied young man. Actually, my body's been pretty badly beaten up. Although superficially I appeared fit and healthy, climbing those stairs safely without incident meant I had to focus hard on my balance, redistributing my weight to my better limbs and grabbing the handrails for dear life, while grimacing at my displeasure that I could no longer disguise my physical disabilities which I just about got away with in my normal subdued walking mode. Except now, the full production might of a BBC's camera crew was literally focused on me, and potentially several million people on prime time TV could be witnessing my precarious one legged hobbling ascent to the lair.

I was palpably relieved when I surmounted the last step and landed on the solid wooden floor in a rather dingy and cavern-like room, despite cameras and bright hot lights apparently everywhere; one of which felt like it was simmering above my head. Of course, there were the five mysterious investors, none of whom I recognised as it was the very first series of the now cult BBC reality business program, Dragons' Den. At that time, very few business people were well known as business matters were considered too dull and serious for prime time TV while business people were tainted with not having particularly attractive personalities in real life. I had no idea what the Dragons did or what to expect. I was told by the production team to aim for the designated cross on the dark wooden floor, and then give the pitch. But I didn't spot it and missed the mark. Immediately, filming was halted and the director instructed me back down the stairs to repeat my entry. That was actually rather unpleasant for me as I am, in fact, quite badly disabled in my right knee. I also felt quite embarrassed as a whole crew of people watched my one legged descent. I did not warn the BBC about my disability as I didn't expect such steep stairs to the studio. In fact, when I saw it, I requested to go up by the lift, but they refused, because as it turned out, climbing up the stairs is part of the scene!

My pitch was for a product which at that time was only an idea: a neat solution which could minimise missed appointments at the NHS. Every year in August there is an annual report about how the NHS is losing around £300 to £500 million pounds because people don't turn up for their appointments, for one reason or another. Most of these so called 'DNAs' (Did Not Attends) are caused by people who genuinely forget, although a significant percentage is due to negligent and selfish people who don't give a damn. Because of my bad knee, I needed to attend hospital quite regularly, and had never missed an appointment unwittingly; except, one day, I opened my drawer to find a paper appointment card with an appointment date which was over a week old. "Bugger" I exclaimed, feeling rather sorry that I had also committed a DNA.

I thought: If only the paper appointment card could have made a beeping noise to warn me of my imminent appointment, I would not have totally forgotten about it. There are many ways to remind someone to do something or attend an imminent appointment such as ringing the patient, sending a mobile phone text message, writing the appointment down in a diary, or using an electronic organiser. But these methods make assumptions about the patients, i.e. they are at home when the reminder calls are made, or they possess and use their mobile phone actively. Therefore, I think the most inclusive way is to make the appointment card beep at you. But a piece of paper won't do that!

Once I got into this mode of reasoning, I had a wonderful Eureka moment. Can you imagine what it might be? Maybe our solutions may even concur, or you might just have come up with an even better idea! I will reveal my invention in a later chapter.

## I want to be normal

For most of my adult life I'm used to walking rather slowly. I walk with a slight limp on most days but limp quite pronouncedly on not so good days. In fact, sometimes I can't walk at all. Friends and even family members usually only see me on my good days, but occasionally they catch me on the bad days. "I twisted my ankle on the kerb", "I fell and hurt my knee" are my usual prepared excuses to deflect their concern. In reality, the reason for my transient and quite frequent signs of disability is due to a lifelong medical condition. I was born with it, but for most of my childhood, I tried to deny it and rebel against it. Because by rebelling, I thought I could beat it, and show that I didn't need the medical intervention to keep me like a 'normal' person. I paid a heavy price for my defiance, as insidiously I became increasingly crippled by it. I do not entirely regret it though, because the unruly and recalcitrant acts of my childhood may have also saved my life too. What is this condition? In the past, I was never able to be fully open about it because of the stigma and fear of prejudice which accompany it. But time and life experiences have taught me that openness makes life far easier. The affliction which I had for so long, that felt so difficult to express, was really just part of the diverse normal human condition. I am now grateful for my illness as it gifted me the motivation and creativity to embrace abnormality – to be an inventor and the author of an eccentric life story.

## 2. ISLANDS AND SEAS

It is not possible to write about one's life without first revealing the life of one's parents. They were sceptical about my decision to write this book, thinking that one would have to be a very important and successful person to consider such a deed. I have not travelled the world, partaken in revolutions or done anything dramatic; however I felt impelled to tell my personal journey because through luck and circumstances it reflects many of the events and challenges of society and changes in the world. I am not sure if my parents were convinced though, as my life has appeared to increasingly veer towards that of an impoverished artist. But their lives have led to my existence, their genes and their actions have very much led to who I am. For this, my journey starts from their journey.

Friday 13th October 1967 was when I emerged in this world in the bustling enclave of Kowloon Hong Kong, delivered in an upper floor flat of a recommended maternity 'mama'. It was a privately run 'birthing centre' popular with immigrants or those with less means; giving birth at the local hospital would have cost several times more. The birth could have been especially precarious given that both my mum and I had an undiagnosed medical condition. Fortunately, despite a difficult birth, there was no crisis, and we went home three days later with support from my great aunt. My dad was not around to help out as he was a seaman working somewhere in the South China Sea. This was a tumultuous time in Hong Kong and the world; the year long Cultural Revolution in mainland China had spread to Hong Kong. Stirred up by the fervour in the North, for over a year, between October 1967 to May 1968, Hong Kong experienced its worst rioting in history, where many people were injured and even killed from violent riots and bombs planted by fervent Communist sympathisers. It must have been an anxious time for my mum who mostly raised me and my older brother alone, as my dad had started his career as a galley boy in the British Merchant Navy ten years earlier. My parents were immigrants to Hong Kong looking for a better life opportunity. They were born in the now famously entrepreneurial city of Wenzhou, about 350 kilometres south of Shanghai. China was a very poor country when they were children. My father's mum died during childbirth when he was just two years old, and he never attended school. He was already working with his fisherman father at around 6 years old, catching fish and shrimps from a rickety boat in the local river and estuary. By any standard, life was hard, but fortunately living in such an underdeveloped area meant their indistinct village was relatively untouched from the Japanese occupation army in nearby Hangzhou and later the civil war which followed the defeat of Japan at the end of the Second World War. My mum was more fortunate, she grew up in a somewhat less arduous surrounding in a traditional and spacious Chinese courtyard house, sharing with her parents, 3 brothers and 2 sisters. It was a very social and neighbourly place to live as there were 13 other families residing in the courtyard compound. Most of the adults were classic Chinese farmers tendering and subsisting from small plots of land. My mum even had a few years of schooling up to the age of 13. When her cloth shoes became too tattered to walk the rough path to school, that became the end of her education. Although academically bright (she claimed she was top of the class!), unable to replace her shoes was good enough reason for her to give up school. At that time, girls from rural villages didn't need to be highly educated as the expectation was to marry a hardworking man who would provide enough food and start a family. Looking at the youthful black and white picture of my mum, in her late teens, dressed in a flowery dress, she was quite pretty and tall, with a strong feminine curve, a kind friendly smile expressed through her high cheekbones, and a classically pale complexion as desired throughout Chinese history. Her face was not classically beautiful as such as she'd inherited the big fat nose of her father and the broad lip of her mum, and her otherwise youthful skin was scarred by acne due to her excessively oily skin which still afflicted her today. Of course, the picture didn't reveal her spotty face, she was really attractive overall, and no doubt would have attracted the attention of local boys. As my mum was not particularly adept in working the field, my dad was introduced to her by her uncle, hoping she would follow him to escape the hardship of farming the land. They were being matched as potential marriage partners. The old pictures I saw were used to net my dad who was then already in Hong Kong. Independent dating was rare or even not acceptable at that time. Chinese society in the 60's was very conservative, but not as repressive as earlier time. My mum could decide if she liked my dad enough to pursue him further. She needn't worry as my dad was quite a handsome young man and stereotypically a hard working Chinese country boy who wanted to move on, to escape the poverty of his village, and to seek out greater opportunity elsewhere.

That 'elsewhere', was the small island of Hong Kong; an intertwined product of East meets West. Hong Kong is one of the world's great cities, and probably recognised as such as early as the 1970's. In the 1950's, Hong Kong was much less well developed, although it was already being recognised as a stable place with economic opportunities, whereas China was just recovering from several decades of internal strife, colonial exploitations, civil war, and conflict with Western powers in the Korean War. For many Chinese, Hong Kong represented a chance to make a better life. There were two major waves of immigration into HK; in the late Forties to early Fifties during the bloody civil war between the Communists and Nationalists, and then the first major Cold War conflict in the Korean War. The latter led to a UN embargo of China and an exodus of established merchants and entrepreneurs from China, especially the Shanghainese who exported themselves and their factories into Hong Kong, which laid the foundation of HK's clothing industry. My great uncles and aunts who settled in HK were from this wave. They escaped from China, not just for economic opportunities, but to avoid the almost continuous trend of instability that has blighted China within their living memory: wars, political chaos and grinding poverty. My dad's arrival in HK represented the second wave, in the second half of the Fifties. He left China mostly for the benefit of greater economic opportunity, although the benefit of hindsight meant he also escaped almost two more decades of disasters of Mao's "Great Leap Forward" and the infamous Cultural Revolution[1]. Ironically, it is often people from the poorer villages with less education who had the most desire to emigrate. Many of my dad's uncles and aunts had already made their way to Hong Kong in the late Forties and early Fifties, while with the exception of one of her uncles, all of my mum's relatives remained in her relatively 'better off' village.

So in 1957, the young seventeen year old fisherman's son followed a number of other young men in his village to a new land. But there was no official route to Hong Kong. Their way was the way of the Snakeheads route, the illegal smuggler gang who organised the journey. It must have been an exciting feeling for a country boy to make such a voyage, probably without fully comprehending the risks. They had to first get to the coast in the southern province of Guangdong; this part of the journey would have taken several days by rough roads and steam powered trains. They were then smuggled into Macau, a sleepy Portuguese colony on the tip of China's coast and 35 miles south west from Hong Kong. The real danger was the trip from Macau to Hong Kong. They rested there for two more nights as the snakeheads prepared and gathered other illegal immigrants for this more treacherous route. Dozens of young men were packed in the hull of a rickety speed boat and sped rapidly on the shark infested South China Sea. My great aunt who sponsored the trip was extremely worried as the night of my dad's departure was overtaken by a severe storm. Without detailed knowledge of his crossing, her relief was palpable as she heard the 17 year old young lad knocking on her door amidst the howling wind and gloomy dark night. He has made it. In those days, once an illegal migrant had made it to Hong Kong, he was entitled to settle, just as tens of thousands of previous migrants had.

## A Chinese immigrant in a foreign Chinese land

My parents' generation represented one of the latest waves of immigration to Hong Kong in the late 50's from mainland China. They were not refugees escaping conflict, but economic migrants escaping poverty and seeking greater opportunity. Most of the indigenous population and early immigrants were from the southern province of Guangdong (formerly known as Canton); hence the predominant language spoken in Hong Kong is Cantonese while English is the reserve of the British and educated elites. There were also immigrant groups from other provinces, most notably the Shanghainese from Zhejiang province who brought with them many of their business skills in the manufacturing industries, which helped lay the early foundation of the textile and other manufacturing industries in Hong Kong. A less well known and smaller minority group also from the Zhejiang province were the people of Wenzhou, a small port city with a sea-trading heritage. Flanked by the sea on its east and surrounded by mountains along its inland boundary, its local geographic isolation gave rise to Wenzhounese being famed in China as exceptionally self-reliant with strong kinship bond. Most curiously, Wenzhounese is the most peculiar and notoriously incomprehensible dialect amongst all the hundreds of dialects spoken in China. The Wenzhou phonetic and grammar is so unique and sounded so weird that an interesting legend and probably true story was the deployment of Wenzhou people to convey sensitive military communication during the Sino-Japan war and up to the Sino-Vietnam war in the late Seventies[2]. Today, Wenzhou is famous for being the cradle of Chinese capitalism.

## From SS Inchearn to Atreus

So, what does an illiterate teenage boy who doesn't speak the common local Chinese language do in Hong Kong? Like many of today's immigrants, the network of relatives already settled provided a support network of help in temporary housing, translation and introductions to job opportunities. Ironically, weakness in the local dominant Chinese language led to a broader outlook in seeking employment. A popular job opportunity for many non-Cantonese speaking Chinese immigrants was to become sailors. Historically, the British merchant fleets were one of the biggest employers of Chinese sailors, with a sometimes bittersweet history going back to the 1870's. In January 1958, just one month past age 18, my dad started work as a galley boy in the SS Inchearn[3], a cargo ship for the Williamson & Co shipping company. An understandable unskilled position for a young man with no formal education, it was a physically demanding job working round the clock to assist in all manners of physical tasks, from assisting in the kitchen, cleaning and moving heavy goods around the ship. Between departure and arrival back to the Hong Kong port, each trip lasted between 4 to 6 months. The accommodation was cramped, shared with other Chinese seaman from different regions of China and Hong Kong. As well as working, it was in this ship that he started his formal education. With the help of a particular patient colleague, he studiously learned to read and write Chinese, as well as learning English too since that was the commanding language of the ship's British bosses. Because his background was so disadvantaged, he felt he needed to demonstrate his ability by being extra diligent and always working his hardest. His diligence and hard work paid off when he became the sous chef within three years in 1961. In 1963, after a year of communication by letters, my dad returned to Wenzhou to marry my mum in order to get the necessary certificate for her to settle in Hong Kong. In 1964, their first child, a son was born; a healthy and boisterous boy. My mum and dad were just 21 and 24 years old respectively. It was an exciting time for the nascent family in the cramped quarter of Kowloon's To Kwa Wan's district. However, financially it was a particularly hard time. My dad was present at the birth of my older brother because he had to miss some sea journeys due to ill health. It was the decade where he suffered persistent bad health due to a lung condition which was not diagnosed at the time. He would experience periodic heavy bouts of coughing that would sometimes even spew blood. Before my parents had their own public housing flat, even my aunts and uncles who put him up in-between his ship duties were anxious about the cause and worried for their own health. The log book's medical records showed he was cleared of infectious diseases such as tuberculosis, so it was most likely internal injuries caused by over exertion during his physical work duties as a galley boy and then as sous chef. This set back his career as he could not keep his chef position with this condition and had to miss some journeys. Frustratingly, he returned to galley duties during the second half of the 60's. I was born in this uncertain period as my dad struggled to regain full health and to feed the extra greedy mouth that was me! There was quite a contrast between my older brother and me. He has very dark skin, whereas I was milky white. He was boisterous and quite difficult as a baby, whereas I was as the Chinese would say a very "gua" (or roughly translated "well behaved") baby. Just as well, as my mum had to raise us mostly as a single mother since my dad was away at sea for most of the year. Just as well even more that I was this well behaved baby, because within a year of my birth, I would present to my innocent young mum deep anxiety and worry. A condition which was undiagnosed at birth had presented itself totally unexpectedly. The doctor had diagnosed a rare blood disorder. It was a shock to my parents and to my mum in particular as she would be the one who would mostly deals with it.

In 1970, my dad had transferred to work in another cargo boat called Atreus. Compared to today's gigantic behemoths, it was a tiddler but elegant 463 feet long vessel built by Newcastle-upon-Tyne's Vickers-Armstrong. It was owned by the famous Blue Funnel Line, which has the signature blue body and black tipped colour funnel. Fortunately, his health had recovered, and he quickly regained the sous chef position and later obtained his only official qualification, the Chief Cook Certificate from the Nautical Catering College of Liverpool. In 1972, he was promoted to Chief Cook and remained so until May 1975 when he left a seafaring career for the solid ground of Liverpool. Despite his decade of poor health in the early years, it humbles me to see that his log book record under 'General conduct' and 'Ability' was stamped with "VERY GOOD" throughout his entire 17 year career at sea.

The decision to work ashore and to immigrate to England via Liverpool is a well-trodden path of many previous generations of Chinese sailors. Although Hong Kong was an increasingly affluent city there was a lot more competition than if he made use of his skills in England. Even though mid 70's England, and Liverpool in particular, were declining fast, it was still one of the most affluent countries in the world; therefore, there was a much greater opportunity for any hard working immigrant to expand their economic opportunity. Even more so, if the immigrant had niche skill where they could stand out and make a living. For most of the Chinese migrants at the time, it was the ubiquitous chip shop and Chinese takeaway! My dad was familiar with Liverpool as the Atreus had last docked there a year earlier, and some of his friends and contacts had made similar journeys. These people were the pioneer migrants who made the latter migrant paths easier. In order to be given right to stay, he was offered a job as counter and kitchen assistant in the local takeaway owned by one of his older Shanghainese ex-colleagues who had settled in Liverpool many years earlier. The pay was just £35 a week with lodgings, less than his previous job, but with a much greater prospect for building the family's future. Thanks to the super strength of pound Sterling, my mum assisted by her part time home knitting job was able to keep the three boys well dressed and fed while I played the older brother to the newest member of the family; my baby brother who was born in 1974.

Like most ambitious and capable Chinese immigrants of his time, my dad did not come to England just to be an employee in someone's shop. The plan was to start his own business, and that opportunity came within a year of his arrival in Liverpool. A run-down chip shop in a council estate in the Lancashire town of Oldham became available. It was by no means certain he could achieve his goal so quickly as the leasehold was a considerable sum of £6000, which was many times his savings. Bank loans were impossible as he had no track record or the skills to draft a business plan. But he did eventually manage to get the funds, from relatives from Hong Kong and America who would often scrape and risk their own savings to help out. This high degree of willingness to lend money between relatives at preferential terms is one of the main reasons for the rapid transformation of many Chinese immigrants who have nothing or very little, to being small business owners. An interesting comparison of my dad's immigration journey is he could get into the UK legally, whereas many Chinese seamen, including my dad's maternal uncle, chose to settle in the USA years earlier by abandoning their ship during port call. They would usually 'jump ship' when they had a place to stay either with friends or relatives who had settled earlier. Meanwhile many would often work as reasonably well paid house servants to well to do American families until they could obtain full citizenship in one of the frequent amnesty calls.

## Memory of early childhood

People are often asked about the very first memory of their life. Often they will recall a particular event at such and such an age, sometimes amazingly at less than one year old! I imagine most people would remember themselves recalling certain vivid events or experiences before they understand the concept of time or indeed about their ages. This event usually then requires verification by another person (usually the parent) who can verify the event corresponding to their recalled time. Without this verification, it is impossible to be certain that the event of that very first memory ever existed. Having verified my first memory with the help of my mum (or was that memory implanted by my mum, into my consciousness from stories recalled to me at a later age?), it appears my first memory corresponds to a scene of people around my infant bed, in a brightly lit room. My mum was possibly there, but the most vivid image was that of another woman who wore a bright white dress and a crescent shaped cap. There was a pounding pain in my forehead. I cannot remember if I had seen myself in the mirror, but I have an image of myself as a big headed infant with a big round hard lump above my eyes, coloured deep purple. It was very sore and sensitive to touch. I was bearing it. It was a huge towering bruise which protruded from my forehead.

My mum had taken me to the Queen Elizabeth hospital in Kowloon, where the initial batch of medical staff was baffled. I was a cute little 11 months infant who a few days earlier had banged my head on the cot, and the bruise just got bigger and bigger. It was only after some further tests that a doctor diagnosed a rare blood disorder. It was "hyut jau bing" as the doctor would have told her in Cantonese, which literally translated as "blood friend disease". Colloquially the doctor would have explained to my mum that I had the 'lau hyut bat ji' disease, which literally translated as 'bleeding would not stop'. In fact, the explanation would have come through to her via translation from a friend who accompanied her. Being a country girl from Wenzhou who had only arrived in HK just four years earlier, she would still be struggling with the local Chinese Cantonese dialect, and adapting to her new home. The diagnosis must have been a big shock to my young mum who had never heard of the condition. Unlike today where most people could learn more about a medical condition via the internet and social media, in those times, people like my mum could only know as much as a doctor had told her. In effect, that meant she knew rather little superficial knowledge. She would be getting to know a lot more about my condition as I grew.

One thing which I wish to clarify from the onset about my condition is that if I cut myself or bang myself hard, the bleeding would not be any faster or any more intense than a normal person. It just bleeds longer as the blood doesn't clot properly, so any wound or injury does not heal as quickly. For small external cuts, this isn't much of an issue, as a simple plaster or a little pressure via a bandage would heal the wound. The main problems faced by people with this condition are internal injuries, in the tissues or in the joints where stopping the bleeding is not so easy. In the case of my big lump, a normal child would probably just have a slight bruise, whereas mine continued to bleed from the injury site until the bleeding ceased due to the internal pressure of the accumulated blood. At the time of diagnosis, my hematoma ('bruise') had probably already reached that state. It was painful, very tender, and vulnerable, but the internal pressure had finally enabled a firm clot to form over the injured site and stop the bleeding. The bruise had stabilised and subsequently started to subside. There is no cure for the condition, and in the Sixties, no readily fixable treatment as is available today. As it started to settle down, there was not much the hospital did for me, except bandaging me up and telling my mum to take extra care of me. Over a period of several uncomfortable weeks, I recovered completely.

Inevitably, I have a succession of memories which revolve about my condition: an ear bleed after having poked a foreign object into it which necessitated several days in hospital; a disabling leg bleed which necessitated piggybacking on my mum's back as she took me across bustling Kowloon to see a doctor in a clinic. I recall that particular visit vividly as we were threatened by a street vagrant carrying a stick while I became aware of the embarrassment of being piggybacked on my mum at my age. I must have been less than 5 years old at the time. Except for the unusual clarity of the above event in my memory, the only other thing I can recall in the first four years of my life is a hazy recollection of walls and wandering through narrow dark alleys and staircases. This scenario makes sense as we were living in those famously cramped and somewhat dilapidated looking buildings of Kowloon's To Kwa Wan and later Tse Wan Shan resettlement districts. Ah, just a bit more, with the privilege of reminiscing intensely in the course of writing, I have also recalled the rare memory of being held by my dad, and jumping naughtily on my parent's bed.

The passing of infancy and toddlerhood coincided with my family leaving old style Kowloon to a brand new estate in Hong Kong Island. There begins my clearest childhood memory of the joyous walk, hand in hand with my parents and older brother as we walked into Wah Kin building containing our new flat in the modern expanding Wah Fu Estate on the south part of Hong Kong Island. According to mum, we moved in 1972 when I was 5, the same age as the Estate, and two years older than the building. We had moved into one of Hong Kong's earliest and biggest public housing estates. It was built to alleviate the population expansion of Hong Kong, partially caused by decades of intense immigration from mainland Chinese immigrants such as my parents. The buildings of Wah Fu were moderately large high rise buildings housing around 400 families per block. Socially, it was a pioneering place as it was the first public housing estate designed under the new town concept providing its 10,000 plus residents[4] a sense of community; it had a fine balance of amenities including schools, shopping malls, restaurants, a public library and most memorable of all, a large wet market selling fresh vegetables, meat and locally caught sea creatures. In the early days of its construction, some people were reluctant or even scared to move there, as the land which it was built on was previously a public cemetery. However, as more people moved in, Wah Fu became popular not just because of its affordable housing, which was built for people with less means, but because it was located in one of the more remote and beautiful locations of Hong Kong, with a beautiful sea view of the South China Sea. All the flats had their own kitchen, toilet, and a balcony which were particularly welcoming during the humid and hot season. Ours was just one bedroom though, so the living room also had a double bunk bed to accommodate me and my elder brother. For a family of five (or even four, as my dad was mostly away) it was by Western standards tiny, no bigger than 330 square feet; but a significant improvement to the cramp and barely minimal Kowloon apartment. Our flat was at the corner of the building, overlooking two primary schools, one Christian, the other Buddhist; beyond it, we see the shimmering of waves of the nearby sea, with the remote Lamma Islands between it and the expanse of the South China Sea. On hot days, our flat was frequently bathed by a gentle sea breeze, whereas on the cooler days, the worst of the weather and wind were shielded by the unforgettable and classical triangular profile of the Lamma Islands. Wah Fu Estate was surrounded by the sea and rugged coastline which also hid Hong Kong's only significant waterfall bay. My favourite childhood adventures were time spent, usually alone, in these beautiful places, being fascinated by the diverse wildlife of this area. As many people since remarked, Wah Fu Estate would not have been built as a public housing project today; its sea views and location are more suited to millionaires than public housing tenants!

Even without the benefit of hindsight, I felt really lucky to have lived in such a beautiful and diverse environment, which, as I grew older provided me with the opportunity to play and explore. Perhaps because of the biodiversity of these surroundings, as I was growing up, I developed a strong interest in nature. I loved observing and catching the incredible variety of wild fauna that were literally within our front door. Some of these were not so pleasant, such as giant cockroaches that hid in the nooks and gaps of the flat, audibly speeding across the floor during the night; and manifesting themselves as dead bodies in the morning caused by the poison baits. From outside through the open balcony, mosquitoes of various kinds would inflict severe itching and red sores after having dined on my blood. For some reason, I recall most vividly, while sitting on the toilet, observing industrious ants wandering purposely on the bathroom's concrete wall, while a pair of them appeared to communicate intensely with each other. I sat there wondering what they could be saying to each other. The bathroom was also the place where a cute looking reptile would occasionally show itself, emerging from some mysterious hiding places in the flat. It was a pale pinkish coloured gecko that could seemingly defy gravity by running across the walls or the ceilings as though gravity didn't apply to it. In the summer season, attracted by the lights, the flat would experience an invasion of hundreds of mayflies. Other times and to my delight, visits from various other exotic insects would often land right inside the flat. And this was just the indoor fauna!

As Wah Fu was built on a steep hill, there was a network of open rain drainage systems running round most of the building blocks. Like the river that attracts so many animals in the African savannah, these drains attract much of the smaller animals that dwell in the dichotomic nature of urban and wild that coexisted in Wah Fu estate. I think I must have started to explore these drains when I was just five years old, initially with supervision, but not much later mostly on my own. Some of these drains were quite steep, whereas the one which I explored most, which was just adjacent to my building block was mildly steep, about 45 degrees. It was fun to climb these drains, but also very exciting, as there I would often see a variety of small animals which I would either observe or catch for fun; like juvenile brown grasshoppers which for some reason I liked to catch a lot, using my bare hands. With hindsight, I think I can attribute much of my hand eye coordination and speed to my early years of trying to catch these insects! Nearby, depending on the season and time, one could find giant African snails crawling around the drain, and on occasion I witnessed life and death struggles between them and an army of giant black ants which frequented the place.

Other common species that frequented the drains included frogs, beetles, different varieties of grasshoppers, dragonflies and finger sized red millipedes that would curl up as one touches them; and if further provoked, will emit a foul smelling darkish red liquid. In the nearby vegetation, there were other interesting insects to be found, such as praying mantis, crickets, and cicadas (I could hear them often, but have only ever seen one alive). For some reason, I particularly treasured catching the "Kam Gui Zhee" beetle[5], a spectacular fluorescent pearl green colour, with robotic like underbellies and legs. I like it a lot perhaps it was because they resembled some of the very popular Japanese transformer characters and mega man children TV programmes of the time. One has also got to be careful as there were quite frequent sightings of wild snakes living close by in less disturbed areas. To this date, my most frightening moment was when a bright green snake with a triangular head (mostly likely a green pit viper) literally whizzed round my leg as it emerged, and then in a flash, returned to the rock crevice. Meanwhile in that instant, I just froze with fear, while my heart rate spontaneously went through the roof. The snake was taken by surprise by my presence and was probably even more startled than I was, as it quickly rushed back to its hole. In between the urban and wild, there were also packs of feral dogs roaming the streets of Wah Fu. There was one pack, apparently led by a Dalmatian like breed, which I particular remember as it looked quite friendly. But they were definitely not approachable, and were at times intimidating when I saw them on the street. In fact, I first saw the most graphic 'survival of the fittest' struggle when I witnessed one particular chilling event of a street cat being chased underneath a parked car directly below my flat, which was then pulled out and ripped apart by these dogs. It took me a long time to forget the curdling cries of the cat as it fought for its life. Apart from this horrid event, my most memorable childhood experience was playing by Wah Fu's rocky coastline, and fishing in its dark blue sea. The coast was just a few minutes' walk from my flat and was even closer to the Buddhist primary school which I attended until I was 10 years old.

The above memories were wonderful, and I felt so lucky to have been brought up in Wah Fu. But my childhood exploration and growing up were never going to be typical.

As I grew older, I became heavier, moving faster and generally carrying more energy. I liked to explore my surroundings like other kids. I would jump up and down over obstacles, and play with other kids from the block; one particularly favourite game of daring being who could jump from the highest steps in our building. I would also frequently climb along the rocks and cliffs of the nearby seaside on my own. I would scrape myself or fall as one does as a child. Most of the time, there would be no immediate problem; I just played on… but quite frequently, usually on the next day, and sometimes maybe just within a few hours from a fall or altercation with a solid object, I would notice parts of my body becoming unwell. Sometimes that's all it was, a transient discomfort; but many times the soreness continued to worsen, until I became debilitated with pain. A lot of the time in those early days, when I was about 5 or 6, I developed big bruises on my lower legs, on the shin, through falls or through bangs on a table or a chair. Over a few days, it would worsen, until, typically, on the third or fourth day it would stabilise, and then over several more days I would see and feel it becoming clearly better, until eventually it would be back to normal.

For relatively small injuries like the above, I usually got better within a week and returned to normality soon afterwards. Ever since the diagnosis in my infancy, it was inevitable that there would be further complications as I grew up. My mum had taken me to the hospital in serious situations before. There, I would get better with time. It just seemed that there was no real treatment other than to rest and let the hurt run its course. In one particular episode where I was debilitated, we passed a private clinic advertising the service of the doctors inside. Instead of going further through the bustling crowd and busy public transport to the hospital, my mum and her supporting friend took me in. The doctor looked and sounded confident and knowledgeable, impressing my mum to take his advice and service to treat me. He prescribed rest as well as some liquid medicines which I took over a period. I recall one bottle in particular containing a milky white liquid which was quite delicious. My mum also kneaded a towel dry from a bowl of very hot water to place on my swellings, with the intention that it would dissolve the hematoma. These combinational treatments appeared to work, with me eventually getting back to normal, even though at times it was painfully slow.

Of course, the complications were due to my blood disorder, 'hyut jau bing'. I have delayed revealing the condition in English because I didn't learn the English word for the condition until much later in my childhood; around eight years old. It was perceived as a long and hard word, because unlike many British colonies of the time, many of the inhabitants of HK were quite ignorant of the English language. Only older people who had a higher level education or were from a high status background would have a good command of English. At primary school level, all communication was done in Chinese with simple basic English taught as a second language. In the Seventies, being able to speak English fluently was perceived to be the reserve of the privileged and the highly educated. So when I did manage to learn and pronounce the word, I felt like a smart ass. It was the hardest and most complicated word that I had learnt so far in my education. Although the condition is rare, and most people are ignorant about it; its name is famous. It is called Haemophilia, and I am a Haemophiliac.

When writing the last sentence I felt a big relief to have finally got these two 'H' words out of my system. For much of my life, I have lived with the condition calmly and contemplatively, but I have also fought it ferociously and foolishly. Haemophilia is intrinsically part of me, but it is a condition which I would prefer to keep secret or even pretend not to acknowledge. Over the years, beside the health difficulties which I have encountered, the 'H' words would in time come to mean stigmatisation, prejudice, shame, anguish and alienation. With so much mixed emotion, I feel much unease in just uttering the word haemophilia or haemophiliac. So much so that even for the sake of writing this book, I would rather use the letter 'H' in place of the word. Readers – please bear with me for my awkwardness; whenever you see the letter 'H', please substitute it with haemophilia or haemophiliac depending on the context.

H is not all negative though, without it, this book would certainly be much less interesting. More importantly, I probably would not have been as mindful or developed the intelligence to look on the bright side or appreciate the simplest goodness of life in and out of adversity. Much of my thinking and actions in life have been informed or shaped by H. It has influenced my thoughts on a wide variety of topics including economics, human rights, the human condition, politics, religion, and innovation. It is with this perspective that I hope some fascinating and provoking thoughts will prevail. Meanwhile before continuing the journey, the next section covers a little more background information about H.

## What is H?

Most simply put, H is a blood disorder where the blood doesn't clot properly because of a missing protein essential in the clotting process. The effect for someone with this condition is if he[6] gets a cut, the blood will not clot properly and so continue to bleed for much longer, depending on how severe the injury is and how severe his deficit of the missing protein factor. The modern treatment is via intravenous replacement of this protein factor, either as when required or prophylactically. The latter treatment option involves regular injection (of 3 or maybe 4 times a week) of the missing protein factor to ensure a severe H has a constant minimal level of the clotting protein, so as to prevent any damaging bleeds occurring in the first place. I would like to emphasise that having H does not mean that the bleeding is any faster or more violent than a normal person, or that the skin or tissue is any less fragile. However, because blood from wound or injury doesn't clot as quickly, the bleeding lasts much longer or does not stop until the appropriate treatment is given. In fact, the greatest affliction of H is not superficial cuts, which can in most cases be treated with just a bit of pressure and a Band-Aid, but internal injuries in the muscle tissues or joints where there is no obvious way to control the bleed. Before modern treatment became widely available (in rich countries) in the late 1970's, such internal bleeds were debilitatingly painful, and crippling. As I mentioned in my own personal experience, some bleeds do cease eventually on their own – but only when the internal pressure caused by the exasperated blood is so great within the joint or tissues, that bleeding ceases. This kind of situation is extremely painful, lasting many days or even weeks. Furthermore, as exasperated blood in joints is very 'corrosive' to the cartilage linings, frequent repeats of such episodes are the main reason that so many older H become disabled from a young age.

    I have the most common form known as classical H (hey, I just love the English language in the way it can romanticise a medical condition!) in which the missing protein is known as factor VIII. The severity of the condition depends on how little factor VIII is produced in the body. In my case, I have less than 1% of normal levels, which is classed as a severe condition. Being severe means one could get spontaneous occurrences of painful bleeds in muscles or joints without apparent causes. Those with 1% and 5% are classified as having moderate H, and those with between 5% and 40% are considered to have a mild form of the condition. People in the latter group may not even be aware of their condition, until they experience some serious trauma or are in need of an operation. H is fortunately a rare disease, only prevalent in 1 in 70,000 of the general population in the UK, and a broadly similar distribution across the world. Because of its relative rarity, most non-specialist doctors are usually quite ignorant of the condition and don't really know how to deal with it.

It is also a famous condition despite its rarity. For some time in the early 20$^{th}$ century, H was commonly known as the 'Royal disease'. Queen Victoria was a carrier, and this disease used to run in several branches of the European Royal family, including the Spanish, German, and most infamously in the Russian branch which ended with the brutal execution of Prince Alexei, a H and the only heir to the throne of the Tsar family in the 1918 Bolshevik revolution. If one looks carefully in old film footage of Prince Alexei, he is often shown limping or being carried along by a servant.

Tragically, H became frequent headline news again from the mid-1980's due to a sequence of contaminations in the treatment of H by three deadly diseases: HIV, HCV and vCJD – three acronyms which send shivers down the spines of many people across the world. They were to cause profound impacts to much of the H-community. For my generation, very few H have escaped all three curses. I was no exception in this regard.

## How do you get H?

I used to be quite sensitive about H being called a disease, as though it is something which is infectious or dirty. I much preferred to call it a condition, although it is just semantics, and assigning H a different label does not change the fact it is one of the classical examples of a hereditary disease so widely documented in literature. I remember quite well how I used to try to understand my condition from old encyclopaedias and medical books bought in school book fairs in my childhood. So let's see if I can explain clearly without going into the details of genetic biology.

All living beings have a genetic blueprint made up of genes, and genes themselves come in pairs, with each gene sitting on a different chromosome [a unit component of a gene]. A fundamental difference between women and men is that women have two X chromosomes while men have one X and one Y chromosome. A male will have H if he has a faulty X chromosome ($X_h$) which causes the condition. If a woman has one faulty X chromosome and a normal X chromosome, the latter tends to exhibit greater influence so that such women are carriers of the condition, but is mostly unaffected by H[7]. Hence, it is very uncommon for a woman to have serious H as that can occur only when two faulty X genes are combined, i.e. when a carrier woman has a baby girl with a man with H. Since males only have one X chromosome, they get H if it is faulty. The diagram[8] overleaf hopefully clarifies these points.

Soon after my diagnosis of H, both my parents and my older brother were tested to determine their status. There was no history of the condition from either side of my parents' family tree. So there were only two reasons; my condition was due to a one-off genetic mutation, or my mum was a carrier either due to her dad being a H, or her mother being a carrier, or that she was the source of the genetic mutation which started the H defective gene tree. As we know now, my mum is a definite carrier[9] as her Factor 8 level is in the mild H range, indicating that one of the two genes responsible for H

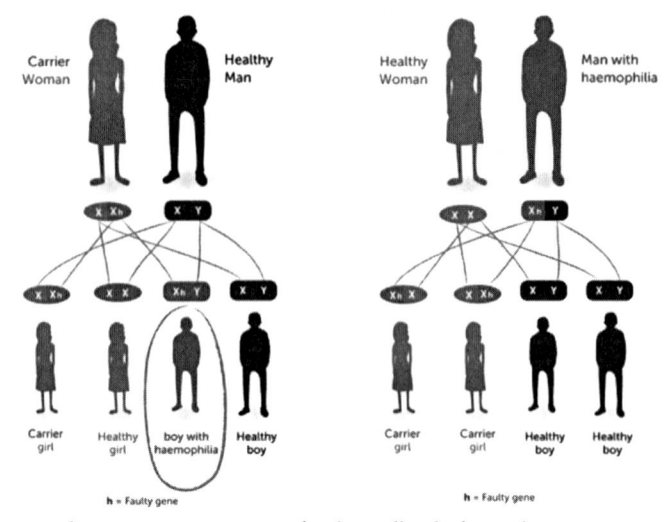

Copyright permission courtesy of Mike Holland of www.haemnet.com

is defective. My maternal grandmother who lived to 86, had three daughters and three sons, all of whom, except my mum, didn't show any signs of carrying the defective H condition. Furthermore, my mum's younger sister had two boys who are free of H. Therefore, together with the fact that my mum's father was extremely healthy throughout his life, living to 90, it appears that the H condition must have started in my mum. These are the definitive conclusions that I can draw now, although at the time of the test in the late 60's, or even much later, I don't think my mum or I really understood the real situation. As a carrier it meant that any male sons of hers would have a 50% chance of being born with H. As it turned out, she had three children who are all sons. Very fortunately both my older and younger brothers are free from H.

## Rebel without a cause?

Being a H, starting school presented my mum with additional worry as the school had to be informed of my special condition, and she had to accept the start of my own exploration in the outside world, and with other children. The pre-primary playschool was located in the same building, on the ground floor near my flat, so going there seemed quite informal and easy going. Most of us remember some elements of their first day in school. I felt anxious but most memorably recall other kids crying as their mums 'abandoned' them in their new settings. I do not recall any significant events at this stage of my education, except my introduction to the teacher (probably the headmaster or admission tutor) as my mum needed to mention my special condition. As expected, none of the teachers had encountered a child with H, and they all had to be assured that other than being a bit more watchful over me, I was just like other kids. On reflection, I would say this was the start of my own awareness that I was not like the other 'normal' kids. I was different and special; but not as I would like it. I wished I was normal, not to stand out as a child that required special treatment or who expected to be treated differently.

I was lucky that my mum never cotton bound me as a kid, partly because she couldn't quite control us three boys, and she over-trusted my diligence to be careful. My dad, like most fathers was much stricter, but because he was a sailor away from home so much he seemed more like a stranger.

Starting primary school changed the dynamics of my life more so than the average kids. There were two primary schools situated just ahead of my home's building, a Catholic denominated school called Precious Blood Primary and a Buddhist denominated school called Chung Wah Primary. The former was the more popular choice because it had a higher academic reputation, and being a Christian school considered better because of the association with the rich and successful Western societies; parents queued before dawn in order to register a place for their young child. My mum didn't bother to do that as she didn't have such a strong view (or perhaps she didn't get up in time), so both my older brother and I went to Buddhist Chung Wah Primary. I recall being slightly disappointed at not getting into the 'better' school, but that was soon completely forgotten as I don't know any better. Anyhow with hindsight there were no memories of bad discipline or poor teaching. On the contrary, what I recalled was tough discipline and lot of crammed learning, made more intense because pupils only attended half time. The school ran two shifts, one in the morning and one in the afternoon, as there were too many pupils on the estate to enjoy a whole day of schooling. The school started with an assembly where all the pupils gathered to hear a speech or prayer from the headmaster or a Buddhist abbot dressed in his traditional colourful robes, which was then followed by us pupils chanting a rather rhythmic Buddhist hymn, sung in Sanskrit! The pupils all wore a simple uniform, and most memorable of all a rather cool looking dark green blazer bearing the school logo of the traditional symbol for Buddhism; an inverted swastika!

Without trying to compare with my later primary schooling experience in England, several things stuck in my mind about the teaching methods in the Hong Kong school. There was a lot of learning material being taught each day; especially the Chinese language lessons where there were many characters to memorise. Furthermore, frequent tests were demanded where pupils had to recite a whole page of a short story, word by word. Similarly, in mathematics, by the age of 10, we were expected to have mastered long division and non-trivial looking algebra. The multiplication table was of course completely ingrained into our brain cells. On top of this, we also learned elementary English lessons taught as a second language. On discipline, the teachers were definitely very serious and harsh.

They were expected to moralise as well as being the authority to obey and respect absolutely. I still remember vividly that each term there would often be pupils wetting the floor in the classroom because they were too frightened to ask to go to the toilet. Corporal punishment was rife, and not just meted out only for bad behaviour; a pupil could be caned either on the palm or the buttocks if homework was missed. In fact, I recall being a recipient of one caning on the palm from a music teacher, for some transgression which I can't remember. The teacher didn't know about my condition, and I was too proud to let her know. Of course, I suffered a lot more than just the momentary sharp pain of the cane[10]. Looking back, I think such corporal punishment was just plain wrong and nasty. In addition to a heavy homework load, even primary school pupils were usually put under a lot of strain. Being a Buddhist denominated school there was clearly an emphasis on Buddhist teaching in the religious school lessons. Other religions were taught as well, particularly Christianity, which was the 'other' big religion beside ancestor worship and some other Chinese deities who were worshipped by some of my neighbours. Perhaps because Buddhism is a more humanistic faith, I can't remember any indoctrination or feeling any strict dogmatic religious conditions being placed on the pupils. An interesting annual demonstration of this easy attitude was that come each Christmas, pupils would habitually send their teachers Christmas cards. Despite not having felt indoctrinated by Buddhism or any other religions, I wonder how much of 'me' has been influenced by the school's religion. A number of Buddha's teachings which I wholeheartedly agree with are concerned with compassion and respect to all living beings. I have certainly developed strong compassion for most life form I encountered. They were mostly insects, small animals like frogs and fishes caught from the nearby sea and fresh water streams, releasing them after playing with it, or looking after them for a while. I would get really angry and cursed other kids who showed cruelty to any animals for the sake of fun. Yet, I did not become vegetarian or show much compassion to the bigger fishes which I caught from line fishing to take home for food. To me, it's this kind of exceptionalism and contradiction that disinterested me from claiming or wanting to belong to a religion. To obey and satisfy all its doctrines seemed impractical and quite impossible. Furthermore, I could not reconcile entirely either with the concept of reincarnation[11]. It's a great concept to deter cruel people acting cruelly to other people or animals, but what about the people or animals who are suffering in the current world? Does it mean they deserve it? Have I been given H because I was a bad person in a previous life and so I deserved it too? The situation is similar with Christianity as well, with the common belief that you will go to hell if you don't believe in God or follow Jesus Christ. I really don't like this absolute edict, which if true is ultimately sadistic and cruel. Therefore, as a young child I was sceptical of religion. Besides, there were more exciting and immediate things to be interested in whilst I was growing up: wildlife, the sea, fishing, stamp collecting and studying. As I learnt more, I was dumbfounded finally to realise the answer to a question I posed to my parents a few years earlier; who is God? I was around 8 or 9 years old at the time, and I learned that Jesus Christ was the son of God, and God apparently was an all-powerful supernatural Being which created the Universe, and all the workings of the world. This amazing revelation did not startle me in as much as Ultraman's (a very popular Japanese children's program at the time about a Superman-like character who battles against monsters and baddies) incredible power and abilities; what really shocked me was that fully grown adults all over the world, from the most illiterate under-privileged peasants, to the most educated world leaders on earth would choose to believe in such an intangible concept as an all powerful Deity! As a kid with limited knowledge, I did not have the confidence to dismiss the belief of so many intelligent and highly educated people; so the logical sceptical question was, 'does God really exist'? If I had just led a normal healthy life, I would probably have just dismissed or ignored any further thought on this question. I may even have become an atheist. But, my H and its unpleasant consequences often led to painful battles. These battles to get back to normality were often long and lonely. There would be one side of me desperately wanting to be well again, thus hoping and praying to be better, while another rational and realistic side of me questioning my sanity unto whom am I praying? There is no doubt that during much of my younger self, I have prayed often; and yet simultaneously felt uneasy of my wishful action.

Most Chinese of my parents' generation and origin have no religion. My parents were no exception having been brought up under the atheism of Chinese communism. However, when in HK, they must have considered or being influenced by my father's aunt and uncle who were devout Christians. I recall accompanying them to church at least a couple of times and feeling slightly intimidated as a small young child in a crowded audience of adults, alternately singing and listening intently to a VIP behind the altar. I was probably between 4 to 6 years old. I remember this man giving a very long speech, mentioning the words "Soeng dai" (God) numerous times. I recall asking my parents about who was "Soeng dai", to which they would just say I was too young to understand. Not satisfied I felt extremely bored, and wishing we could leave soon.

By the time we arrived at the Wah Fu Estate, my great aunt and her family had emigrated to the US and my parents seemed no longer interested in organised religion. Nevertheless, they had kept a framed picture of Jesus in the living room. As my dad was mostly away working on the high seas, I recall my mum would often say a prayer to the picture seeking comfort or best wishes as she gallantly raised her difficult boys; my older brother being particularly boisterous and rebellious, while I could be too adventurous for someone with special medical needs, and my much younger toddler brother requiring constant attention. Furthermore, being a recent immigrant she had few friends or close relatives to turn to for extra support. And of course there was no social security, so finance was also very tight despite her working from home knitting bags after bags of jumpers to supplement the family income.

Perhaps it was within this less than ideal supportive environment that I developed protectiveness towards my mum. That is if I hurt myself I would try to hide it, hoping it would recover by itself. I didn't want to worry or burden my mum unnecessarily. For a severe H like me, this usually meant hiding pain and discomfort over periods of a week or two for a relatively 'small' trauma such as a 'minor' muscle or joint bleed in areas of my body which can be rested readily. Despite being a severe H, I usually recover naturally via a combination of rest, and self-healing via increased internal pressure and the tiny level of Factor 8 working to stop the bleed. I'd put quotes around the words 'small' and 'minor' above because that was how I perceived those injuries at the time. As a stubborn young kid born and growing up with the condition since birth, I didn't know any better. It was pain and inconvenience which I came to accept as normal. However, with the benefit of hindsight, these 'small' and 'minor' injuries could have easily turned considerably worse. Indeed, many times it did get a lot worse and I could no longer hide the condition. Then my mum realised, and she would take me to see the doctor from the private clinic. I still recall the comfort I felt when entering his office noting his impressive Canadian medical degree certificate and professional demeanour. All over the world, professionals such as doctors automatically command great respect from almost everyone. Even more so if he or she has a Western education as that is generally perceived to be superior. Therefore, it was only natural that my mum and I had complete respect and confidence for his advice and his prescribed oral medications which apparently made me better. However, bearing in mind what we now know about how my condition should be treated, it appears that whatever medication was prescribed was probably ineffective. As far as I know from current knowledge, I am not aware of any oral medication that can effectively stop bleeding. Perhaps they were just anti-inflammatories which reduced the swelling a bit faster. To this date, I do not know if he had deliberately kept me as his patient to earn extra income, or whether he was genuinely ignorant about the best course of action. He should have referred my mum to the haematology department of a main hospital for proper treatment, where I would have been offered replacement Factor 8 which had been available by the mid-70's in Hong Kong[12].

This ignorance and my stubborn attitude of trying to hide and fight my condition without treatment had a profound effect on my psychology and future wellbeing. Not receiving the proper treatment meant I endured numerous episodes of painful and debilitating bleeds, usually endured alone without my mum realising it; hoping the episode would resolve itself eventually. My mum usually failed to spot such early episodes because most bleeds were within muscle tissues or in joints. These bleeds, unlike superficial bruising are usually very difficult to spot initially. Using an example of a 'minor' trauma episode in my knee joint as a child I would feel the following experience which had several distinct phases: 1) an aura of feeling something is not quite right. There is no pain yet, just feeling a discomfort as though it could get worse. There could be no physical symptoms at all at this stage, although as time went by the suspect area could feel warmer than usual. There may also be a slight puffiness at this stage. I would feel anxious, but also hesitant, and had wishful thinking that this was just a bit of strain and tiredness. If I took particular precautions and immediately rested myself for a couple of days or more, then there was an occasional chance the bad feeling would go away and I would be back to normal. 2) If it was indeed a bleed, as usually turned out to be the case, and it hadn't been treated or gotten better on its own, then the problem area would become clearly hot to touch and be visibly swollen. There would be clear signs of discomfort and pain. I could probably still try to walk normally if I wanted to, but this would unambiguously induce deep discomfort or pain. Usually, my mum would discover my situation at this stage as I would start to walk with a limp even when I thought I was successfully hiding the condition. Another clue was that my activity levels would drop noticeably from normal. There is now an overwhelming sense of anxiety and worry of what might happen next. If the bleed is not treated, it would then proceed to the next more severe phase 3) which would be obvious debilitation due to significant swelling and increased pressure in the joint or muscle tissue, loss of range and serious pain, especially if any strain or weight is put on it.

Having been discovered, there is a big relief as I no longer had to put on an act and could allow myself to rest properly, get treated with the probable placebo medication, and be comforted by regular soothing applications of hot towels from my mum, which apparently helped to reduce the swelling[13].

With the resilience of youth, and the subsequent bed rest I would regularly recover from the phase 2 and even stage 3 situations before the bleed worsened. The recovery would take from one week to several weeks depending on the severity and area of the bleed. Mentally, the journey from the first sign of trouble to eventual recovery was dominated by anxiety; the anxiety of the bleeds getting worse, of waking in the morning in a worse state than the night before, anxiety of burdening my mum, and anxiety of falling behind at school. When the condition turned for the better, there was the anxiety that it could revert to bad again. Only when the condition had reassuringly stabilised with a reduction in pain and swelling would I feel relief. When it continued to improve, then there is an optimistic feeling of looking forward to being normal again. When it's almost back to normality, the original source of my problem is no longer my focus; I felt a great relief of freedom and joy of not being bound to the sick bed.

What I have just described are sensations and emotions which are not that exceptional; most people have experienced and can recall a number of painful or very painful periods in their life caused by sporting injuries or accidents. The big difference is even at aged 8 or less, I felt I had probably already endured more physical adversity and health related anxiety than most adults in their entire life. I started to enquire about fairness, noting that my brothers or other kids grew up without any of my 'hassles'. Compared to most kids, I had more time to contemplate as I was frequently lying on my bed to recuperate. I thought about God and Jesus as I started to learn about religions in school. Who are they? Why are they so popular with adults and spoken with such reverence? Are those miracle bible stories true and could I believe in them to help me recover during my all too frequent lonely and anxious battles? My thoughts on these questions were just superficial as I was too young, too impatient and too ignorant to delve further. I was interested to learn more, but had no inclination just to believe or follow the practice before I understood these legendary, and what seemed to me at the time, mythical figures from ancient times.

Surely, modern science and medicine should be my prayer, not some mysterious unprovable supernatural being. Well, a traumatic event would finally push me back into the world of modern medicine. At aged 8, in the evening, with no warning, I felt a flash of wetness in my pant, and as I put my hand down to investigate, to my shock and horror, my hand was soaked with frightening redness of fresh blood. There was no hiding this emergency, I told my mum immediately. I recall standing up and then immediately fainting, falling backwards, my head hitting the solid vinyl floor of the living room. I regained consciousness almost immediately noting the bloody floor around my trousers, and the frightened look of my mum's face. An ambulance was called, and I was rushed to Queen Mary Hospital, one of the top public hospitals in Hong Kong that was only a couple of miles from my home in Wah Fu estate.

For the first time in my life[14] I was given Factor 8 concentrates to treat this drastic bleed. It had a calming effect as the doctor injected into my vein a syringe-full of a cool clear liquid containing the Factor 8 medicine. After a number of years fighting off the H condition without proper effective medical treatment; receiving proper treatment within the hospital environment felt so much more reassuring, even though the terrifying bleeding that was inside my abdomen was initially only gradually diminished. For the next few days, I could feel gurgling of fresh blood continue to build up inside me, which was released when it couldn't be restrained any longer. As the blood was very red, it was most likely due to some perforation or cut of the intestinal region. Several whole blood transfusions were needed to compensate the blood loss. The bleeding finally ceased after a week of daily treatments. To this date I never fully discovered the cause or its exact nature. Although I had seen a few doctors by then in my life, this hospital stay transformed the way of how my H would be treated in the future.

I never saw the private doctor who had apparently over a number of years been prescribing me with (probably) useless medicine. Instead, I was now under the care of consultant Dr Anita Li, at the haematology department of the Queen Mary hospital. As a consultant specialist in a major hospital who had successfully cured my traumatic bleed, both my mum and I were truly in awe of her. As I recall, her seniority was matched by her tall and elegant poised stature. Also she was friendly, always speaking kindly and patiently to my mum who could not always communicate freely in Cantonese. She informed us that any future bleeds must be treated as soon as possible to prevent greater hurt and damage. Finally, for the first time, H didn't seem like a life sentence as it could be treated effectively. The treatment was free too as Hong Kong had developed much of its health service based on the National Health Service of England, where treatment based on needs were free[15]. Just as well, as I grew older, some of my bleeds in the muscles and joints became more serious. This was probably due to my heavier weight and higher speed of collision or fall, as well as being more active. The availability of Factor 8 at my local hospital meant I could get treated effectively whenever I wanted. However, the availability of new treatment did not alter my old habit and self-inflicting mentality of hiding my pain from my mum and hoping it would recover by itself. I resisted treatment until I was found out or realised that it wasn't going to get better on its own. Hence, when I did eventually go to the hospital for treatment, I was usually in a much worse state than if I had presented myself a few days earlier. However, I felt amazing every time I was given the injection of Factor 8. There was almost an instantaneous like healing as the usual throbbing or intensely sharp pains of the afflicted site calmed down almost immediately. And after a day or two, the painful symptoms almost disappeared and I would feel almost normal in just a couple of days. This was even truer for bleeds where I allowed myself to be treated almost as soon as they were detected. The uncomfortable feeling or mild painful symptoms associated with an early bleed literally just disappeared within an hour or two of the treatment.

Contrast this with my previous existence, where the initially mild symptoms would usually develop into a more painful and debilitating situation, where with luck the bleed may eventually run its course and then I could experience the precarious relief of recovering. Depending on the severity of the cycle, from first symptom to recovery could take between several days to several weeks, or even months to resolve entirely. Recalling this earlier time, there was really hardly a day when I was totally pain or restriction free. The application of replacement Factor 8 was like a miracle; when I got treated by it, I felt like Superman, as though I could do anything, and invincible to any hurt that could be thrown at me.

So why did I continue to be so stubborn about getting treated more quickly, especially as I had started to experience some particular unpleasant and debilitating bleeds in my hips and ankles? The original reasons mentioned earlier remained true irrespective of treatment method: a reluctance to bother and worry my mum; a stupid denial of my abnormality; and the self-belief that I could get better by myself as I had most of my life until I discovered Factor 8 therapy. This culture of resistance and rebelling against H had been ingrained in me, so even the benefit of the replacement therapy didn't change my attitude and habit. This resistance was further reinforced by an over-concern not to miss school, and a misleading public duty to be self-reliant. The latter came about through my awareness that Factor 8 drug was very expensive, so I felt I ought to minimise my treatment despite the fact that early treatment would minimise potentially more severe development of the injury. To Dr Li's credit, I only recall her kind and gentle advice on the importance of early treatment in order to prevent further worsening of the condition, rather than any complaint about the cost of treatment. However, I remember one particular incident where another doctor severely and sarcastically lectured my mum on the cost of the treatment. This incident was to have a powerful influence on my wellbeing and subconsciousness. It further impacted on me that not only did I cause burden to my family, I was also a burden to the State. There can't be any doubt this knowledge reinforced my reluctance to be treated as quickly as I ought to. In fact, even today, my hesitation to administer treatment regularly still resonates with me – although the main reason now is a mild case of needle phobia.

A young child is probably at their most carefree stage in life if they are fortunate enough to be brought up in a loving and peaceful environment. When old enough to attend primary school, the social dynamics change considerably. For me, I became more aware of my difference as I was advised to miss out most PE lessons. Time appeared to run especially slowly as my classmates engaged in sporting and fun physical activities whilst I was forced to watch. To be frank, I didn't mind missing PE except I didn't like the regular reminder of my difference and the perception of how other pupils thought of me. It was in these moments that I often reflected on if only I could become normal. Having said that, my social interaction with the other kids was quite average. I had a few good mates like most other kids, and couldn't recall any notable prejudices due to the differences caused by my H condition.

Academically, the Hong Kong primary education environment is very competitive and high-pressured with all subjects taught to a very high level. It is easy to fall behind if one misses school for just a day. Furthermore, the importance of education is deeply ingrained and respected even in very young children. It is probably a cliché to say that Chinese culture venerates education, and Chinese kids in general knew this almost instinctively. I for one was no exception in this respect, perhaps even more so than most because I knew my career options were restricted. Hence, my rebellious streak to defy my medical condition as if I didn't have H ironically led me to attend school as much as I could, even when I should really have been going to the hospital for treatment or resting at home. The result was that I would often be in a classroom bearing pain and anxiety while at the same time trying to concentrate on the lesson. One particular bad incident was of a school trip to the newly opened sports stadium. I desperately didn't want to miss it, but at the same time I knew that something was not quite right in my hip at the beginning of the trip. As we arrived there and I eventually got to sit down in the stadium seat, I wished I could just stay seated as the pain became more intense with each standing ovation. It was made considerably worse because on the way to the stadium I had insisted on trying to walk normally; to pretend there was nothing wrong with me, even though I was clearly limping; I was just too proud to show my disability or to bother the teacher. The hours in the stadium felt like an eternity, so it was a big relief when I eventually got back home; and eventually to the hospital where I had to stay for over a week until I recovered. Inevitably, I ended up missing more school than most, which in turn made me more anxious for fear of falling behind. Fortunately I enjoyed learning and studying, so even though I was constantly anxious about falling behind, I was able to keep up despite having to miss chunks of schooling every now and then.

As a H growing up with the complication of health issues, I felt more mature than other kids of my age. Perhaps it was the early hospitalisations, experiencing homesickness, away from mum, anxiety, hope, and boredom. Experiencing the institution of hospital and interaction with older adults and meeting other young patients all increased the range of my social interaction. I recall being surrounded by a group of doctors (the consultant and her students) who were talking in a mixture of Chinese and English, feeling impressed and slightly intimidated by their presence. I particularly enjoyed seeing students and junior doctors who tended to listen to me and give me answers which I liked to hear. Of course, they were keen to learn about H probably a lot more than trying to have a conversation with an 8 year old. Even so, they appeared impressed by the size of my head and seemingly enjoyed the hilarious attempts to measure its circumference. I probably learned the English word for H in one of these hospital stays, and developed a desire that one day I would become a doctor too, just like them and maybe finding a cure for my condition. Another pleasant experience was the hospital food, especially the rice dishes with slices of tasty chicken on bones. It was probably nothing special, except for the rare taste of 'eating out', which was a special treat then.

There were also some negative experiences in the hospital; the auxiliary support staff and even the nurses didn't generally convey compassion or a caring attitude. Many gave the impression that they were just there to do a job rather than also to appreciate the satisfaction of making the patients feel better. It was common amongst relatives of patients (including my mum) to give a small gift to incentivise them to be kinder and more caring. Perhaps it was their low pay coupled with the long hours, but still, a kind attitude towards patients shouldn't have to be paid for! The other bad memory was the difficult attempts to administer Factor 8 into my vein. Although I had good veins in my hands and arms, they were often overused during hospital stays with numerous blood tests and treatments, making successive injection harder. Often, several pokes and attempts at different veins were needed to succeed. In these situations, it became clear that some doctors and nurses were much better than others; I remember well with disgust one particular hopeless junior doctor who failed to get through after more than five attempts! On other occasions, veins from my ankles and even my neck were used.

I recall most of the fellow patients appeared to be suffering from afflictions that were much more serious than mine. I was usually in hospital looking like I simply had a physical injury, whereas the other kids looked jaundiced and sick, most commonly caused by kidney failure or some other blood disease such as leukaemia. Most poignantly I once saw a dead baby and a toddler being wrapped up by nurses while peeking through a gap in the not entirely drawn curtains. As a kid, there is a naïve belief that death is for the very old or those afflicted with severe injuries, but witnessing lifelessness on these young bodies made me very sad as I pondered on the unfairness of life while reflecting my own situation. My first emergency in-patient experience could well have been fatal had modern medicine not been available. Knowing there was now a treatment for the illnesses caused by H must have been a gigantic relief for my mum. Perhaps, I might even have developed an exaggerated sense of security that any illness which befell me could be instantly fixed by the marvel of modern medicine. Beside the afflictions caused by my severe H, I was a healthy boy rarely taken ill by any other childhood illnesses. But not long after my injections of the life-saving Factor 8 treatment started, I developed jaundice; my skin and the whites of my eyes had turned a sickly yellow. I was diagnosed with acute hepatitis B, an infectious and potentially deadly liver virus disease. I was once again hospitalised, but this time for something other than H!

I cannot remember much of this stay in particular, but when I was released from hospital the recuperation involved months of a strict diet avoiding fat and some other foods, and a daily oral intake of various medicines. My mum looked after me well, and I was a good patient adhering strictly to the diet and medication. I made an excellent recovery and completely cleared the virus. But the fact behind the cause of my infection was more disturbing. The Hepatitis B virus was transmitted by the Factor 8 treatment. It was inevitable that this would happen because Factor 8 is manufactured from donated blood, in a process sourced from a pool of up to 20,000 blood donors. At the time, the Factor 8 was sourced from the US, where many of the blood samples were obtained from rough prisoners who gave their blood for money. The infection of Factor 8 by blood born viruses was virtually guaranteed. Yet nothing was done about it by the manufacturers, or even the doctors who prescribed it. From my understanding now, there was a patently lax attitude towards the safety of Factor 8; the medical community generally just accepted that all the risks were worth the benefits. When I made my full recovery, I didn't think much of it as well; I just thought that was part of life; get sick, get treatment, get better. Of course, as we all know now, in the course of the following decades, many H sufferers and their families were to face more deadlier viruses, resulting in the greatest medical disasters of recent times.

## From Hong Kong to England

In 1977, after over a year of running his own fish & chip shop, my dad decided it was the right time to apply for the family to join him. I remember that year well as it coincided with the Queen's Silver Jubilee. There was much talk at the school of the Queen's prospective visit to HK that year, and I was one of the many eager pupils who ordered the commemorate brochure of the event. I was most disappointed when the school failed to secure enough brochures for everyone who had ordered it. To us kids, the Queen was the ultimate authority and the biggest star in the world. In our naïve view, she was not just The Queen of England, but the head of state who held the ultimate power; and yet she appeared so benign and kind. Therefore there was much respect for her, even if we didn't really understand the limited role of the monarchy. We were of course also taught the history of Hong Kong and how Great Britain wrestled it from China in the 19th century through force and violence in the Opium wars. Even though the teaching of this history was all very matter of fact, there was some ambivalence towards the British for most working class Hong Kong Chinese. I would say most of the students (including myself) felt shame and injustice of this history; with the clarification that the shame refers to the shame of the weakness of China and the Chinese people to be bullied and coerced by a much smaller country over a century earlier. At the same time we respected a lot of the laissez-faire but capable and benevolent administration of the British, which combined with the industrious Chinese made Hong Kong one of the most desirable and stable places to live in Asia. For many HK Chinese, their positive feelings (or at least indifference) towards their colonial master was further enhanced when they saw or witnessed the backwardness of mainland China.

I saw China too during this period; biennially between 1972 and 1977, my parents would take the family back to Wenzhou to visit relatives. These were memorable trips; almost 900 miles from HK, the way to Wenzhou was an arduous journey then. First, all travellers from HK to China carried lots of goods; gifts and consumables that were simply unavailable in drab Maoist China. At the time, these goods were put in a bag and then carried by hand or were balanced on a bamboo rod carried on the shoulder[16]. As some of the trips were without my dad, my older brother was needed to help out with the load. I only carried the lighter bags as any bleeds on this journey would be troublesome for everyone. At the HK-China border were the Chinese custom guards who always looked so miserable and unwelcoming, as though they despised people, for whatever reason. With their green official uniform they looked like soldiers, and inevitably the passage through the border was always tense and uneasy. Once we passed the border, it felt like another world. Our regular clothes suddenly stood out compared with the uniformity of drab blue or green garments of the mainland Chinese. The space seemed more expansive as there were hardly any tall buildings or cars clogging up the roads; except for bicycles and three wheeler taxis whizzing around the expanse of roads connecting Guangdong railway station. Although people's lives seemed more relaxed than hectic HK, strangers seemed more distant. Three particularly ill-mannered behaviours that shocked me were: staring, where some people just seemed oblivious to their own actions, staring at us like we were exotic animals; spitting, where people on the street almost seemed like they were competing to see who could spit furthest; and senseless queue jumping, which really made me angry and despair at their uncivilised manners.

Leaving Guangdong, we travelled north to Wenzhou via the main long distance transport of the time, on a powerful steam locomotive. It was an exciting journey as we wound through the tranquil Chinese countryside over one and sometimes two nights. As the train stopped at the stations, farmers and local villagers would offer chopped sweet sugar cane through the windows. This natural snack beats chocolate any day! We also particularly liked the simple fresh food prepared in the train cooked on charcoal burners. The rice was dry but had a particularly smoky aroma which complemented the fresh fried vegetables and light dishes that were also available. As we arrived at the train station closest to my parents' villages, I noticed a particularly peculiar feature of the people: it seemed to me all the people of Wenzhou had really nice glowing and rosy cheeks just like in those ubiquitous propaganda posters. Perhaps it was the contrast caused by the cooler climate, or it might just have been my imagination. We travelled there on a minibus for several more hours through further undeveloped and equally rough roads. Arriving at the village we were greeted by a string of relatives and accompaniment of their children and friends. As children, it was wonderful to be doted on by our uncles, aunts and grandparents as we only got to see them so infrequently. However, we didn't really communicate properly because we didn't speak each other's language! Even though my initial mother tongue was Wenzhou dialect Chinese, being HK based, I grew up in a predominately Cantonese language environment, so I could speak Cantonese fluently but could only understand my parents' home dialect. So even at home, my mum would speak to me in one language (Wenzhonese) and I would reply to her in another (Cantonese). Having said that, by the end of our visit, both my elder brother and I became much more fluent at speaking my parents' home dialect. At that time, many of my relatives were subsistence farmers, and the condition inside their houses was extremely basic. Water was from a well, the toilet was a bucket, cooking was made on a clay furnace fired by coke charcoal, no fridge and no television. At night, the roads were pitch black and in the morning I woke to crows of roosters, and eerily distant howling which sounded like wolves (there probably were wolves in the mountains nearby). There were some bonuses though, the houses were large, which meant playing hide and seek was possible, and one of my aunts even had a pet dog which I befriended. Even though the food served was basic, we had great appetite whilst there, which I think was due to the freshness of the organic ingredients. The experience of my brother and me with other children was not so good though. When we first arrived, lots of the other children persistently annoyed us; gangs of them would follow us everywhere we went, even peeking through our ground floor bed room windows, pointing and making fun of us; we were like aliens from outer space which deserved intense observation. They would make fun of our hair and clothes, particularly the flared trousers which we wore at the time (I don't blame them for that!). As is often the case of children poking fun at one another, it led to name calling and other provocations such as looking down and calling us soft city boys from HK, which was ironic as city people tend to look down on country people. To our frustration, our language barrier didn't help either. Eventually, our differences were resolved when my brother got into a bruising fight with one of our cousins! I mentioned this particular story because it was the first time that I experienced bullying from a group of children through prejudice and being different.

Of course, prejudice is often two ways. In fact many HK born adults of that era, including friends and relatives of mine had never even visited mainland China. They were not necessarily against China, but just held sufficiently negative attitudes of China for them not to visit. I am very glad that my parents didn't have this prejudice feelings or I would never have had the opportunity to broaden my experience of China in the 1970's. Indeed, I was there during the infamous Cultural Revolution period. Fortunately, as far as I am aware, my family's villages didn't experience much of the upheaval, although whilst in town I did see several open trucks carrying standing 'prisoners', with shaming propaganda placards on their backs. Even then in my innocent youth, I felt a chill about their fate.

In these visits up to 1977, China and the conditions of my kin didn't seem to improve much. There was enough food to eat for everyone, but even basic amenities such as flush toilets or TV's seemed like a distant dream. Indeed, I was lucky that I didn't have any major bleeds during my visit, or I might not have survived. It was nice to get to know my grandparents, uncles, aunts and cousins, and to have experienced the different cultures and living conditions, but China the country, badly disappointed me. On returning home to busy HK, I would often ask myself "Why is China so poor and backward?" Why can't she be advanced like Japan? Or like better off Taiwan which was also ruled by people of Chinese descent. Is it because of Mao's bad governance or Communism? Why….? Why…?

My feelings about home were about to change drastically – my dad had began the process of bringing mum and us three boys to England. My dad had settled in well, managing his own catering business in the northern former mill town of Oldham, near Manchester. There should have been no problems in getting us to the UK through the usual immigration procedures administered by the Home Office. But, there was me, a child with a chronic medical condition called H, a condition which is rare but expensive to treat. I became aware that there was fear that this fact could prevent the application from being granted or perhaps that I would have to be left behind. Whether this fear was justified or not, there was a period of deep worry about my condition's effect on the application, and indeed ultimately about my future. There was speculation that I may initially have to be left behind in Hong Kong while my mum and brothers made their new life in England. Of course, I am not saying my parents would have abandoned me, but for practical purposes I would have been left to live with relatives who were probably as equally anxious at the prospects of looking after me; a child with a (to them) mysterious and troublesome medical condition. This was a very worrying prospect for a 10 year old kid.

Despite the uncertainty over the application, my elder brother and I frequently pondered about our prospective new homeland; England[17], the imperial colonial master of HK. As kids, we learned about the British Empire and naively rated its greatness by how large and extensive it was. Although the British Empire was much reduced by the 70's, the sunset view of the British Empire was still too recent to dim Britain as a glorious country with sunrise opportunities for many aspiring emigrants around the world. As Chinese, we also knew the greatness of China, its great inventions, great size and record enduring civilisation that had massively influenced world history... but that glory belonged to the distant past. Compared to the contemporary China we knew, Britain was clearly at the top and we were really looking forward to moving to such a great country. There was also great expectation at being in a Western country; the image of the white man's society as pervades through TV and other popular culture is one of wealth, high sophistication, beautiful tranquil environment where everyone is ultra-polite and lives in luxuriously big houses. This impression was further supported by the luxurious lifestyle of a relatively small and elite group of British and Westerners living in HK at the time.

By the time the Queen visited HK in late November 1977, my parents had received the news from the British Home Office. We had all been granted visas to stay in the UK. The fear over my condition was just that; great news for the family and personally a huge relief. My new home would soon be England, the land of the brave as it is literally translated from Chinese. What would it be like?

# 3. LIFE IN A NORTHERN TOWN

Our date to leave HK was to be on the 25th February 1978. I can't recall much of the preparation for this seminal journey, but I can remember very well my feeling the week before the flight date. Instead of high excitement and a sense of trepidation, I was filled with dread for my health. A week before the flight I had fallen heavily on my buttocks as I played on my brother's skateboard. The skateboard had slipped under me as I tried to ride it over a kerb. As usual, I did not tell my mum. Instead, I hoped that nothing would come of it. That was, of course not the case, the upper buttock area where I had hit the floor bruised up and was becoming sorer by the day. Having landed so heavily, I became seriously concerned that I might have sustained internal injuries. There was real fear that at any time a gush of blood would rush out from my bottom. This was really one of my more stupid acts of bravado, a totally unnecessary risk which could most unhappily disrupt the family's most important date. The first few days were very worrying as I experienced a worsening of the pain and bruising. On the fourth night I went to bed praying all would be fine the next day, or I had better get to the hospital. As usual in this situation I did not sleep well; but when I got up, I sensed the condition had turned, it felt better than the night before. Looks like I had been lucky. All was going to be fine.

The day had arrived; mum dressed us in our best clothes and shoes, ready to leave HK for good and soon to be reunited with dad. We were all excited; first time flying, first time outside of HK/China, and soon first time in a totally new culture where big nosed tall white people are not exotic, and speaking English is the norm. There was the funny but pleasant 'butterfly in the stomach' feeling as the plane climbed into the air from HK's famous Tai Kak airport. The brilliance of the famous HK night lights easily shone through into my memory despite sitting in the middle of the plane. There was a little sadness to see HK behind us, but that was overwhelmed by the prospect of meeting dad and experiencing a new country. Two interesting anecdotes stuck in my mind during this flight; a HK TV personality famous for his travel adventures (equivalent to Britain's Alan Whicker) was in the same standard class cabin as us. And if my memory serves me right, our plane refuelled in Kabul or Tehran – then just another exotic but safe en-route stopover.

Before coming to the UK, I was taught that the English people have a great sense of humour and like to joke a lot. I didn't know what this really meant, but as we landed in Gatwick and proceeded through immigration, I noticed the immigration staff were good natured and courteous despite our nervousness of not being able to understand or speak English. I recall one officer laughing with her colleague as they seemed to remark at my 'obviously' big rounded head. I didn't mind at all as the laughs and smiles really soothed all of us and immediately confirmed the famous reputation of English people having a great sense of humour. Finally, we all got to greet my dad at the other side of the barrier. It had been over 2 years since we had last seen him back in HK, so it was a great joy to us all. After years of long distance separation, the whole family was finally in one place. Up until then, my brothers and I had been almost entirely raised by my mum, having experienced most of the disadvantages of being in a single parent family. As we sat in the back of the hired Vauxhall estate car, I felt proud that my dad could drive and even had his own car[18] at home; a privileged position for most people in HK. There were a lot of questions during the trip. Most of these questions were directed to my mum though because we were too shy to talk to dad directly. When will we arrive? What's our place like? What does the shop sell? Are people friendly? I knew we were going to Oldham, but that didn't mean much to me then[19]. The long motorway journey impressed upon me several things about England: it's really big, and there are a lot of red brick houses in the distance with spans of green fields in-between. Most interestingly, my most memorable part of the journey was the weather, which I only discovered later is a favourite talking topic of the British people. It was a typical dreary day, cloudy mostly with occasional bursts of rain, but not particularly cold for February. Nothing memorable, except when occasionally the Sun was no longer obscured by the cloud, its rays hitting my arms through the car windows felt more warm and welcoming than I had ever noticed before. I think it must have just been a psychological effect of the plane travel rather than any real intrinsic power of the winter English sunshine.

## Initial perception

After over three hours of mostly motorway driving, we finally arrived in Oldham. The landscape had changed noticeably, from flat roads to a hilly incline up a minor road; with one side full of red terraces and a stone wall on much of the other side. Then there were some impressive large houses dotted on both sides, just before we turned left to what looked like a housing estate with a mixture of terrace houses and box shaped low level flats. And just before I realised it, we had arrived at the tarmac courtyard outside the back entrance to my dad's Fish and Chip shop. The courtyard was flanked by a tall red brick wall, reminiscent of a prison! Like so many similar units of the period, our home was an apartment directly above the shop. This was not the idyllic picturesque image of the English home that I had imagined, but I was glad to finally have arrived home. A narrow staircase at the back of the shop opened into a corridor that led to a living room, two bedrooms, a bathroom and a kitchen; each room had a skylight as the building was constructed with a flat roof design. I was impressed by the size of our apartment which was at least three times bigger than the HK flat. But I was also disappointed that my perceived image of living in a lush house didn't materialise; there was no carpet in the flat, and most disappointingly, the sparsely decorated living room only had a black and white television set. I thought everyone in the West had a colour TV. However, dad entertained us with quality food and drink, bringing upstairs from the shop a plateful of deliciously prepared chopped chicken. As we treated ourselves freely with it, my thoughts told me that we must be in a much better position as we could only afford to eat chicken meat occasionally in the past. As we settled ourselves in for a good rest, my dad went downstairs to the shop to prepare its opening for the evening.

The first few weeks in our new home felt really strange and lonely. England was just so quiet! People were so discreet; with the exception of the occasional car or person visiting the shops in our block, most people seemed to be indoors, out of sight and sound. And on Sunday, the entire world as I knew it just seemed to be frozen in time, eerily silent with nothing to do. It would take a while to re-adjust from the bustling sound and noise of ultra-crowded HK. The weather was decidedly weird too; one moment nice sunny skies, but within a matter of minutes clouds everywhere. And often there was rain and sunshine simultaneously while mini hurricane-like gusts seemed the norm rather than an exception. Their howling presence further amplified by the flat roof of the apartment. For most of the time in this early period, my brothers and I were extremely homesick and bored. We wondered if we would ever adapt to this 'quietness' and apparently boring place. The easy solace of television didn't help much either. Why wasn't there any Chinese HK made programmes? I grew up watching TV from HK, US, UK (unknowingly to me at the time, my favourite children's programme Rupert Bear was produced in England) and Japan. There were only 3 channels: BBC1, BBC2 and ITV, and during the daytime, the only thing they showed was either the 'Test Card', bearded men giving 'Open University' lectures, or some seemingly mind blowingly dull programmes like 'Emmerdale farm' and 'One man and his dog'. I even lost interest in my favourite children's programmes; Sesame Street and Rupert Bear because they were not in Chinese, speaking in their original language which I didn't yet understand. It is funny how I loved watching these programmes when they were dubbed in Chinese but just could not connect with them when I couldn't understand the language. Compared to HK, and because I didn't understand much English or its culture, I found British television excruciatingly boring and some of its situational comedy childish[20]. Fortunately, there were exceptions that most of the family enjoyed and looked forward to, where understanding the language was not so crucial, such as wildlife documentaries like ITV's Survival, BBC's wildlife programme by David Attenborough, action films like James Bond, the slapstick comedy of Benny Hill and hilarious 'Carry on' films, some general prize-quiz shows such as 'Family fortunes' and 'Top of the Pops'. The latter was particularly interesting because it was in this programme that we saw cutting edge performances and outrageous style which completely overwhelmed the dull and uninspiring performances of Canto pop-stars. There was one other television revelation; I loved the TV adverts in the UK which were so much more entertaining and creative than HK's boring matter-of-fact adverts.

Dad's shop was called King's Chippy and was situated amongst a handful of other shops in the block; there was a Spa supermarket, a newsagent, a laundry and a betting shop. They were the only retail units serving the area of Sholver; a large housing estate on the edge of a steep hillside overlooking open countryside and the Pennine Hills. Without realising at the time, Sholver was also built in the second half of the Sixties, just like my previous home at Wah Fu Estate. Both had superlatives of being in a beautiful area with Sholver estate variously given the honour of being the highest (about 1000 feet above sea level) council estate in Britain. Like Wah Fu, it had a high element of social housing, although this time the Kam family had moved up the social scale to that of a shop owner. King's Chippy was predominately an English Chippy selling traditional fish and chips, local favourite Hollands pies and Yorkshire puddings, plus a select menu of classic Chinese takeaway dishes. In writing this book, I returned recently to Sholver to check out my old home and was appalled to see the neglect and general crumbling facades of the area. At the time I moved into Sholver, it was only about a decade old, and my memory of the Estate's condition was quite decent indeed. The houses, although basic, were not built in a cramped way, and they were surrounded by nice countryside. There was not much crime or any graffiti issues, and most importantly for the family, the vast majority of adult customers were nice and polite. But there were issues. As far as I was aware we were the only Chinese family on the estate. In the beginning, as a new immigrant kid, this felt very strange to me. When in Wah Fu, I was one of many other Chinese kids in a Chinese society, but now my brothers and I suddenly stuck out as the only Chinese kids among a virtually entirely white English council estate. Although most adults were decent and none threatening, the kids and teenagers of the estate were a completely different story. It was virtually guaranteed that as soon as we stepped outside our house we would be subjected to a series of jibes and racist taunts from the local kids. Some were relatively innocent and innocuous like standing straight, putting their hands together in a Buddhist praying manner and then bowing whilst muttering "Ah So" or some other ridiculous impersonation of how Chinese might sounds to them. Others liked fraying and flapping their hands and feet whilst making ridiculous Kung Fu noises. Bruce Lee and Kung Fu movies were big hits at the time, so almost all kids seemed to think that all Chinese are Kung Fu masters! Although annoying, I wasn't too bothered by these actions alone which were just silly mockery. But unfortunately we had to endure much more than that. A very common affront to us was them pulling their eyes with their fingers until it contorted to a pair of fine slits while chanting chinky, chinky… ah so, chinky etc… as well as a particular frequent rhythmic jibe we heard a lot "Chinese, Japanese, don't forget to wash your knees". Sometimes they looked and sounded comical when it happened, but there was no doubt even to them that this was not just some joking banter between friends. They did it to provoke or make fun of us. As my brothers and I couldn't speak much English at the time, we would often just ignore them. However, sometimes we couldn't just shrug them off, the insults were more severe and personal like telling us to "go back to your own country", "we don't like chinks", "Chinaman, go home" or "fuck off chinky". I would then normally retaliate by hurling the same profanity back, or raising the newly learned 'V' finger sign, and sometimes baffling them with the HK equivalent: thumb sticking out between the first and second fingers of a clinched fist. Most of these verbal abuses were almost always hurled from a distance, so the situation normally petered out as we walked away. It seemed totally unreasonable to me why people wanted to be so offensive, when there was no provocation. Well actually not; just being foreign and Chinese was enough provocation for these brats! Unable to speak or comprehend English well, I felt incredibly frustrated not being able to give them a piece of my mind. Later on, I developed a new way to reply to their abuse; I would copy their insults literally pulling my eyes into slits and call them chinky, or whatever racial terms they used.

## Steep learning curve

After almost two months of acclimatising to our not entirely friendly environment, my brothers and I were ready to go to our new school. Since mum didn't know English, the parental duty had now reversed in some areas; it was my dad who sorted out the schools. I would attend Derker Junior School on Afghan street, in the Derker district of Oldham, a couple of miles down from Sholver estate. Going to a new school is always going to be a nervous time for any child, but it was inevitably a lot more worrisome for an immigrant child of different culture and language. My biggest concern was the inability to understand or speak to fellow pupils and teachers. There was one particular phrase I learned to say in case there was this communication problem; "I don't understand". So if anyone said anything to me which I couldn't make out, I would just mouth "I don't understand", "sorry, I don't understand, please say it again", or "I don't understand, please speak slowly". In fact, I recall feeling quite proud that I had learned these long English phrases! There was a common believe at the time that young children's minds were so flexible that they would be able to speak the language within a few months of arriving in a new country. So optimistically, I assumed I would speak fluently within a few months of starting school. The second worry was etiquette and manners; it had always been assumed that white people were exceedingly polite and imbued with polite etiquette, such as how to hold a knife and fork, how to eat properly with the mouth closed, saying hello, goodbye, please and thank you a lot etc.; stereotypes that I assumed were universal. I made sure I learned these before starting school. The biggest anxiety of all was how could I make new friends when I couldn't speak much English? Would they be like those name callers on the Estate, or misbehave like those kids in "Grange Hill"[21]? There was, of course, one other area of anxiety that neither my brothers or few other immigrants would experience, the school had to be told of my H condition. Teachers would exempt me from PE and some other activities as a precaution. Like before, other kids would be told about my condition. How would they think of me? Just like before in HK, I didn't like it. I wanted to be discreet, be normal and be just like any other kid, but there was no choice.

When the first day of school was finally arrived I desperately wished there was another Chinese kid in the classroom; someone who could translate for me or guide me into the system. If not, maybe there would be other ethnic minority kids who could empathise with my situation. Dad drove me to the school on my first day. I was introduced to my class teacher, an old man with white hair and a moustache, who looked and sounded very much like the stereotypical image of a kind English grandpa. He muttered something to me which

I didn't understand, but I felt less nervous now that I was in the school, instead of anticipating it. I looked around at the other pupils in the class. All white kids except for one black kid, called Geoff[22]. I was disappointed that there weren't any Chinese kids in the class who might have been able to support me. But it didn't matter; I had mastered the phrase "I don't understand". Most of the other pupils were very curious of me, being the only Chinese kid they had probably ever met. Our communications were awkward and often embarrassing as we frequently had to find new expressions to understand each other. But the good thing with children of that age is that they are often more patient and willing to play along with words or sign expressions. I had one particular advantage over some of the other immigrant kids in my situation; play breaks were optional for me because of my H. I thought it unnecessary, but being shy and not particularly keen on football, I didn't mind staying in while being absorbed by a series of wildlife books around the classroom. Sometimes, a buddy was assigned to me, which was fine, as I got more chance to practice English.

Inevitably, many pupils introduced themselves by making funny Kung Fu noises, pretending to be a Buddhist monk or being Japanese with their favourite words "Ah-so!" They were nearly all harmless. Actually, I had by then already taken an amateur interest in Kung Fu, influenced by those amazingly cool fights in martial art movies[23], and the possibility of needing to defend myself from troublemakers like those who made serious racial slurs in the Estate. But it was the fascination with martial arts of the other students that almost caused me serious grief in the first week of school. Whilst waiting in an outdoor assembly, Geoff who had befriended me, without warning punched me in my stomach. It was a miscommunication; he thought I was preparing to take a punch from him to demonstrate my Kung Fu strengths. It's not often one gets hit in the abdomen, especially unprepared, and that simultaneous sharp, hollow and deep pain was excruciating. I put on a brave face, but was actually terrified of internal injury. As before, I didn't tell my parents and hoped I would be ok.

Well, after a weekend of worry, I went to school as usual, relieved and apparently without any sustained injury. I had been lucky and made sure to be extra cautious from rough play or similar antics which were uncommon in HK schools. There was an interesting background to my relationship with Geoff besides this rather unfortunate incident. 1970's HK was pretty much mono-ethnic, so it was very rare to have social encounters with almost any other race. Therefore, just as some of the white pupils were particularly interested to know me, I was particularly interested to know a black friend. Another thing I felt was we might develop a bond which related to us being the only ethnic pupils in the class. Indeed, we got along fine, and he was my occasional buddy; but my expectation of a special bond due to our shared challenges of being from an ethnic minority background never materialised. The big difference was Geoff was born in England, brought up in England, spoke English natively, probably had one white parent and so he didn't share my difficulties of language, encountering new culture, wondering if I would ever fit in, or even dare I say, from observation, experience the level of casual racism that I encountered.

After settling in, I usually took the double decker bus to school. Here, I discovered a huge disappointment as the bus travelled through the hilly incline of Ripponden road. The stone wall which previously hid the scenery behind it had become apparent to me from the top deck of the bus. It revealed more spans of green valley and in the distance some beautiful rolling hills. It was nice, but I could not see the sea!

Throughout my life, I had been surrounded by the sea, and therefore had naively imagined all along that the sea was always nearby. There was another disappointment, on playing outside in the fields, even in the remote places and reservoirs around Sholver, the fields seemed devoid of insect life. Where were the grasshoppers, dragonflies, crickets and various kinds of butterflies and beetles? Similarly, the reservoir seemed devoid of fish or aquatic life (actually there were plenty of fish, but they were just not as visible, exotic or abundant as the fresh water streams back in Wah Fu). The beautiful vast landscape of England just couldn't match the biodiversity of tiny HK!

Another new phenomenon I discovered on the journey was the vocal and physical expressionism of the pupils. On the bus journey, which passed an abattoir where one could often see hanged carcasses, virtually all the pupils would audibly make that "ugh" or "eww" sounds while simultaneously contorting their faces in disgust. At the time, I thought this behaviour rather amusing. They must have seen it many times on the same journey, so why the repetitive surprise and silly expressions?[24] A bunch of Chinese kids in HK in the same situation would not have made such an event a sharing activity. If there was something new or unusual, they might just talk about it between friends.

There was another observation from my daily bus journeys which surprised me. The journeys were frequently quite boisterous, with things being thrown about, kids shouting and making a nuisance. This kind of ill-discipline was just not tolerated or expected in HK. But some of these expressions are quite charming too, like kids waving at pedestrians from the school bus, or when they saw a cute pet, expressing that extremely empathic "awww" sound. I don't think there is an equivalent of this sound in Chinese, except uttering the word 'cute' to express a similar sentiment. There seemed to be many perks in the English education system which HK lacked, like free milk in the morning, free text books and even free stationary. I also noticed many more books in the classrooms, and the regular book fairs in the school, selling new and bargain used books from local libraries. I developed a love for books from going to these fairs and buying bargain books for pennies. A lot of the kids also got free lunch through a voucher scheme, and PE included fortnightly swimming lessons. The latter was particularly poignant for me because my exclusion of PE included swimming, where I looked on from the side-lines introspectively amidst the fun and laughter of the other children. Also, compared to Hong Kong the classroom size was much smaller, and the teachers much more approachable and not strict at all. As a result the whole classroom atmosphere was very easy going; in HK, all students had to sit upright, intent on listening to the teacher, and only spoke when asked to. However, there are downsides to the UK system, like frequent disruptions during lessons. But the discipline in HK schools was probably too much, too soon; whereas in the UK I thought the classes were too laid back; when was I going to learn something new? In fact, I was really shocked by the low level maths that was still being taught. Yes, the class was still doing basic arithmetic at aged 10, while long division and simple algebra had yet to be taught. As for other subjects, I can remember thinking to myself that I knew a lot more Chinese words than they know English words. In my limited understanding of the British education systems at the time, I wondered how come England had such a famed reputation in science and education?

I really treasured learning, and liked to study, whereas I got the impression that many of my fellow class mates didn't take their studies seriously. I thought them spoilt not to realise how lucky they were to have such excellent facilities. Although disappointed with zero maths learning, I didn't mind too much because mastering English and adapting to the new environment was quite challenging enough. In particular, there was the persistent challenge of being subjected to racism by a minority of the pupils. As an adult recalling my experience, it seems hard for me to describe a young innocent 10 or 11 year old child as a nasty racist bully. They would insult me with different degrees of racist undertones. As I described earlier, I could tolerate silliness, or even ignorance based mockery, but it was much harder for my dignity to ignore persistent insults with racist overtones. Unlike the abuse in the estate which was usually shouted at a distance, the situation in the school occurred in confined space in the classroom or in the playground. Hence it was not long before I encountered these unsavoury abuses in the playground. There was one older boy in particular, who made a habit of being nasty whenever he saw me. One day I faced up to him. The commotion led to a group of other kids witnessing our argument, and soon it escalated to talk of fighting it out in the toilets. This was not an escalation situation I desired. But I felt embarrassed and wrong to back down, or he would be in his arrogant and bullying way even more, feeling more justified to look down and racially abuse me in the future. Unintentionally, I found myself standing up against an older and physically bigger boy, possibly about to get into a dire fight. There was no doubt this boy had an aggressive and nasty streak in him, or he would not have taken pleasure in being a bully. I was not a small kid either, and despite missing out PE in school, I was quite fit due to my competitive nature to improve myself in play and study. In fact, I had been teaching myself several forms of martial arts from books bought from a Manchester Chinatown's bookshop. I was particularly flexible with my body, with my legs capable of doing some fancy kicking tricks, including powerful forward and high side kicks which reached above my head.

The walk to the nearby toilets may have taken just a few seconds, but my mind was racing with a multitude of thoughts and emotions. My H condition stood firmly in my background thoughts as I feared the consequences of being hurt, and then hurt more badly. However, at the immediate forefront of my thoughts was how to fight him. I was sure that I could beat him very quickly if I resorted to one of my favourite show-off moves; a very fast and cool high kick with my leg. But I quickly dismissed using my legs as I thought these moves were too powerful and stood a chance of severely hurting him. The adrenaline now rushing through my body appeared to make me light, but also weaker at the same time. I had to focus hard to overcome its debilitating effect. He had now positioned himself in a classic boxing pose. We were separated by a few feet, ready to fight. Having decided against using my legs, my mind was racing for an alternative. Crazily, I recalled the 'madman' style, which involved rapidly spinning both arms whilst moving towards to the opponent. The idea being that the fast rotating arms act as an extension which protects one's head and body while simultaneously battering the opponent. The theory was put into practice; I think I must have hit him a few times before somehow I found my neck around his arm. Crap! He'd got me in a head lock; I was now staring at the concrete floor, spinning and flashes of urinals, desperately trying to free myself from his lock. The adrenaline which previously had almost frozen me seemed to come into its own as I battled against his head lock and attempted blows. That struggle seemed like an eternity at the time, with anxious thoughts racing through my mind… how and when would the fight end? Then suddenly I found myself on the floor. I don't know if I had wriggled free or he had thrown me to the ground. The liberating feeling may have just lasted less than a second before I saw and felt his shoe connecting fully with my face! Probably instinctively, I managed to jump up quickly before he could kick me again. At this stage, other pupils in the crowd intervened and prevented a further round of fighting. I was pumped high and shocked by the viciousness of his kick but was really glad the fighting had been stopped. I quickly checked my wounds from his kick, which had miraculously only caused a small split to my lips as well as a little bruising around them. The bleeding was slight and it stopped after a bit of applied pressure with toilet tissue paper. I was led away by some of my classmates and acquaintances, and I presume he was by his. When I went home, I lied about having had a little accident. And fortunately, despite the fight and the vicious kick to the mouth, I didn't need any treatment.

On my return to school, the classmates who witnessed or heard about my fight hailed me like a hero. I was surprised because I thought I had lost, or at best it was a draw. My opponent had clearly got the better of me. But they considered I had won, because he had apparently wept right after the fight, whereas I had remained stoic. Admittedly it was a close call as my emotions were sky high too. I noticed that some of the kids which had previously been less than friendly no longer called me names, and some even wanted to get to know me. That was a nice feeling. I had shut up an aggressive bully, and been conferred prestige that deterred further provocation. But, I was worried that the defeated lad would want revenge. The street culture of Hong Kong with influence from the gangster culture of the Triads in the Seventies would have demanded some sort of revenge. My anxiety was unnecessary though; soon after the fight, this previously unruly English lad came to acknowledge our fight and offered his friendship to me. I was pleasantly surprised by this drastic about-turn gesture, although I was hesitant initially; why make all this trouble in the first place, and then after being beaten in a fight which he caused in the first place, would he want to be my buddy? This English kids' culture was really weird! Nevertheless, I was really impressed by this 'forget and make-up' attitude, followed by an uninhibited show of respect for one's former enemy. On the other-hand, I was shocked by the completely uninhibited way of fighting here, by kicking me in the face while I was down on the ground. I was certain that had I not got up in time, and if he'd the opportunity, he would have kicked me in the head with all his might, without any thought of the consequence. In fact, from subsequent fights I have seen in streets and several times in dad's chip shop, it seems when people fight over here, they just fight with no inhibition or thought of avoiding serious head injury. In HK, I recall there is a particular ethic ingrained in the society that even violent fights have rules of conduct e.g. one must not kick people in the head which, from my observation of news reports, even Triad thugs who wanted to inflict violence on their victims would avoid doing. I hoped I would never need to fight again, but it seemed to me that if I had to, I would ensure the safety of my own wellbeing before that of my foe. One other lesson, avoid being head-locked in a fight!

Unfortunately this was not my only fight on this playground; there was to be one other physical resolution just over one year later. But before this altercation, I was to experience the toughest fight of my life.

## Futile defiance

By the time of my 11th birthday, I had been in England for over seven months. The sudden change in culture and having to adapt to the new environment had been challenging and stressful. Most immigrants understood this, but my H condition further complicated my adaptation. Dad had earlier registered me with the nearest medical centre which took care of H patients; the Royal Manchester Children's Hospital in Salford. It was no more than 20 miles from home, but compared to the distance I was used to previously, it seemed like a long way away. Dad was a new driver, so getting there first time took much longer as he struggled to find the hospital. Once there, I admired the gracious Victorian building's entrance, its heavy doors and then the long and relatively dim corridors leading to different wards and hospital departments. The H-department was about midway along the corridor, in a more modern development to the right of the original building. I noticed the waiting rooms had walls adorned with child inspired artworks and tables with books and comics. This was quite a contrast to the purely clinical environments of HK hospitals I visited. There, I met H consultant Dr Evans and H-specialist nurse Sister Shaw, both of whom were characterised by their very friendly exposition and big smiles. Dr Evans was tall and self-assured while Sister Shaw conveyed warmth and kindness. Although we didn't really communicate much, I felt reassured that this part of my life was at least sorted out for now. And at the back of my mind I hoped I wouldn't need to see them that often. Indeed, by my next birthday, I had avoided the need for treatment for almost 8 months, despite one vicious school fight, and several other potentially serious mishaps. Of course, there must have been at least a handful of other painful bleeds which I could have avoided had I been more ready to seek treatment. Instead, I had developed a masochistic desire to defy my H, to show the world I could live normally without special medication. I even thought about telling other people with H of my lessons for defeating the condition. In fact, I took particular pride and blessing that I was not that 'abnormal' really – because according to my special medical card, I had Factor 8 level of 4%, which classified me as only a 'moderate' H (actually, later tests confirmed that my Factor 8 level is of less than 1% putting me in the category of having severe H). This illusion of who I was, supported by my early apparently 'successful' defiance of H was soon to be spectacularly shattered by the most foolhardy and deliberate act of defiance.

After enduring so much racist name calling and then my first school fight, my interest in martial arts had increased even more. It was not just for defence, because I knew even then the best course of defence was to avoid violent confrontation in the first place. I became more interested still because it was an outlet of fitness and exercise training which I could purposely learn and practice by myself, when other popular sports were excluded from me. Furthermore, I really admired its elegance and creative motion as demonstrated in the popular Kung Fu movies of the time. In a world where pockets of prejudice and ignorance were common, I felt particularly proud that this great Chinese cultural art form had become so popular and admired globally. But challenging and strenuous martial art training was probably not the best hobby to take seriously by a H-boy. Indeed, it all began with a bit of innocent training; a flying kick and a somersault which went slightly awry. I had over done it and aggravated my right hip.

The exercise had initiated a bleed in the muscles in my right hip region. It was in one of my targeted[25] areas, which had suffered a number of bleeds before, so it had become vulnerable. My response like so many times before was to hope for the best; that it would settle down and then get better on its own. For this to have a chance of working, I would need to be completely disciplined in resting it. Of course it would be better still to tell my parents, seek immediate treatment in hospital, and take complete rest. But I didn't do any of that. I was quite precious about not missing school, so in my attempt to be normal, I continued to attend it, despite the discomfort of walking to the bus stop and roughly 15 minutes of a bouncy bus ride. In my pride not to be noticed or be quizzed by parents, teachers or friends, I continued to force myself to walk in as normal a gait as possible. This was very punishing and stupid as it was probably the most aggravating and hurtful action I could have done, as I was then putting extraordinary stress on already swelling and flaming tissue. The situation unsurprisingly worsened the next day, but yet I continued my pretence, and hoped that the situation would turn the corner for the better. Incredulously, my act of defiance and madness continued into the third day at school. I remember that day well as I forced myself to school. Each step to the bus and every jolt of the bus ride felt like a hammer blow. I was desperate to get to the classroom where I could just sit motionless… but there was first the school assembly. As was normal for the assembly, pupils sat on the ground, legs crossed as they listened to the teachers. I recall forcing myself into that position despite the then intensely tensed, inflamed and swollen hip. The pain I endured in those minutes was just completely beyond anything that I have ever felt. This most profoundly excruciating pain was then combined with the most inner anxious laden thoughts of my life. I almost felt like fainting there and then. I wish I had, but in my most defiant and delirious state of mind, I managed, despite hellish agony, to push myself up into the upright position at the end of the assembly, and then hobble back into the classroom. A bitter regret is why no one in authority in the school could see through the very serious state which I was in? My pride and defiance (and frankly mad stupidity) meant I did not seek help from the teachers or my parents for the rest of the school day. It was the longest day of my life. Instead of conceding to the reality of the grievous situation, I fought and hobbled my way onto the school bus to get back home. It was a 15 minute journey which seemed like an eternity as every slight bump was translated into pure exquisite agony. Finally, I made it back home, sneaking past my parents who were working downstairs in the shop, I laid down on my bed immediately to 'recuperate'; completely exhausted from the most gruelling day of my life. Despite the trauma of the day, even then I still had a slither of hope that maybe tomorrow I would be better. I denied my parents the knowledge of the seriousness of my condition. There was no need to interrupt my parents' evening shift in the shop when I could see the doctor in the morning instead.

There was no escaping hospital this time. By the time my parents got me to the hospital, I could hardly move at all. There was now more than the feeling of pain, a numbness had started to spread into my right leg. I was immediately hospitalised and given plasma cryoprecipitate[26] (which has a high concentration of the key clotting Factor 8) to arrest the bleeding. Unlike previous treatments, there was no magical speedy recovery this time. My denial and defiance of H had led to the most traumatic injury of my life. The initial small bleed in the hip had escalated to a major bleed in the ilio psos muscles, a large set of muscle tissues which runs from the back, round the hip and the thigh. This set of muscles also wraps around the major femoral nerve which they were now compressing. Despite weeks of twice daily infusions of the cryoprecipitate, the intensely inflamed and painful area of the hip region barely stabilised, while even more worryingly the loss of sensation travelled to the tip of my toes, intensifying day by day, until they went completely numb, bar a wave of frequent shooting pain. Worse still, my brain was no longer connected to my entire leg. No matter how hard I concentrated, there was just no response, not even a wiggle of the toes. The whole of my right leg had become lifeless and totally paralysed. Furthermore, the psos muscle injury and the nerve damage had caused the leg to be bent in a right-angled poise. It was not possible to simply straighten it because of the severe contracture of the gravely traumatised muscle. I could only lie on my back as even sitting up would have pulled at the muscles. For the first time in my medical experience, the doctors and nurses could not offer me a definite prognosis which I wanted to hear; like: yes the feeling will return soon, you will be able to straighten the leg again, and yes you will be able to walk normally again. None of the doctors or nurses could know for sure, nor could they offer any assurances of my future.

There was some positivity from the shocking circumstance which I found myself. Although not thinking in the superlative at the time, I was being cared for by one of the best H centres in the country, in one of the best children's' hospital in the world. To my great surprise, all the hospital staff I encountered; cleaners, doctors, porters, nurses, were just so polite, kind and positively nice to me and everyone in general. This was a great contrast to my hospital experience in HK where I remember little genuine friendliness, and in some cases, appalling attitudes. It's a shame that I have to put HK down in this particular aspect, but the reality, in my experience at least, was public services in HK society in that era were much less civilised and developed. However, my pleasant surprise of the nice people in the hospital wasn't just between two countries; it was between the cultures of the NHS and the restricted experience of my immediate surroundings in Oldham. The latter consisting of good decent people I encountered in school, streets and dad's shop, but there were also significant encounters with drunken behaviour and blatant racism which was alien to my previous existence in HK. This negative experience was what stuck out until I was hospitalised, in a totally different environment where the culture of the staff seemed to be entirely benevolent and loving! Of course, not everyone were pure angels, and no doubt some in private may have held unsavoury views, but that didn't matter overall because the NHS culture I experienced was care and cure, and this dominated and united the people working within it.

Of the many great people who looked after me in the hospital, two great individuals stand out; Sister Alex Shaw and junior nurse Ceulia Chin. Nurse Chin was introduced to me and my parents on the day of my admission. Of Malaysian Chinese origin who spoke Cantonese, she was requested by the doctor as an interpreter. Although working in a different ward, she was often called to break down the communication barrier. More than that, almost every day after duty, she would visit me to see how I was, and to discuss my anxiety and experience of the day; I can be quite talkative with the right person, so quite frequently I can drag on the conversation for some time, discussing all sorts of matters. Perhaps, she found my chat quite engaging for an 11 year old kid, or perhaps she was just being kind, either way, for her to spend that extra time with me most days after her work was a truly generous act of compassion and selflessness. I would have felt so much more isolated and vulnerable had she not been there[27]. Sister Shaw was another angel. She had the sunniest, warmest and most caring faces of all the nurses I can remember. She was responsible for treating me most days; her appearance at the ward, with her warm smile, while carrying the prepared treatment in the tray was always reassuring. She was brilliant with the needle, and I never had any anxiety when she was around.

The progress of my recovery was excruciatingly slow. Unlike previous bleeds which were dealt with quickly and effectively with a few doses of Factor 8 treatment, I had now incurred serious damage to the femoral nerve due to prolonged internal compression caused by the swollen tissue in the confined hip region. A plaster of paris was moulded to my entire leg to minimise further stress in the hip region, and to prevent further contraction in my already right-angled bent knee. Then every week, the old cast was removed; the bent knee was straightened up as much as possible while keeping my back straight; a new plaster of paris was then cast for this position. The objective was to enable my paralysed leg to incrementally return to a straight leg position, which was much more preferable than a permanently bent leg. In the early weeks, my knee could at most only be unbent by a rate of about 10 degrees per week. It was progress, but the waiting to see how much my knee would eventually straighten up was with much foreboding because my leg was still completely paralysed, and as the plaster was being replaced I noticed the muscles of my leg had atrophied to the skinniness of a starving child.

After about two months in hospital, the original swelling and trauma to the psos muscles around my hip had dissipated, finally releasing the devastating pressure on the femoral nerve. With concentration, my big toe could once again be commanded to move and tingling feelings had started to return to my leg. The previously severed disconnection had began to be replaced by random and sometimes intense sharp pinprick type pains. My skinny paralysed leg had almost been straightened back to normal. The eerily numb skin on my thigh had started to feel itchiness, made worse by weeks of being encased in plaster-casts. The discomfort was encouraging as it signalled a positive turn from the state of total paralysis.

During this time, I had plenty of time to contemplate about my situation. In the first couple of weeks in the hospital, in the midst of intense pain and worsening paralysis, I felt compelled to make prayers to soothe my desperate situation. I don't remember who I prayed to in particular, except I only know the legends of Jesus, Buddha, Mary and God Almighty. I might well have prayed to them all; it didn't matter who they were since all of them were supposed to be benevolent and supernatural. However, as soon as my condition improved, despite its snail rate of progress, my compulsion for praying naturally subsided. The dominance of my rational side superseded the urge for benevolent supernatural comfort and intervention. I don't think people should expect help from God all the time – I thanked the Lord anyway despite the return of my scepticism.

## Learning to walk

After more than two months of being confined to a bed, my leg could finally be straightened again, without pulling at my back muscles. I could now sit up properly, rest on a chair, or use a wheelchair to the bathroom. This was a wonderful relief as I was beginning to develop bed sores and had become very bored of being stuck in the same place for such a long time. The plaster had now been replaced by a straight leg splint. It had a split along its length, so the leg could be taken in or out of its support anytime. Although there was now a cacophony of feelings coming back to my leg, it was still entirely paralysed except for a few toes; the rest of my right leg was still just a dead weight and as recalcitrant as ever. To move myself or adjust my position, I had to bring both of my hands under and just above the knee, lifting it up and shifting it along. I began to see a new set of health professionals for the first time in my life; the physiotherapists. Until then, I had always taken for granted that all I needed for full recovery was my replacement clotting factor. Physiotherapists are experts in human muscle- skeleton interaction, offering skilful muscular and skeletal manipulation as well as advice on a range of exercises to improve the relevant physical body functions. If I had not sustained nerve damage, their challenge would simply be instructing me with a range of exercises to rebuild the wasted muscles and regain full range of movement. But, I could not do these crucial exercises while my leg was paralysed! Fortunately, there was a glimmer of hope that the nerve damage had been making a gradual recovery. In the weeks my leg had been plastered, my toes were the only part that I could see and feel; they had been the last region to go numb, and the first region to show signs of recovery. By concentrating on them over this period, slowly and surely more toes, and eventually the whole foot, came back under my mental control. The upper leg region was less responsive though, but with the help and encouragement from the physios, and what felt like sheer determination and concentration, I was finally able to twitch my quad muscles. It was very frustrating though as the mild twitch was just that; the intense focusing of my mind to move the leg only correlated with a barely visible contraction of the quadriceps. Still, this was a breakthrough which enabled the physios to guide me on a set of exercises to keep the quad alive and to build up its strength over time.

With much trepidation, I spent my first UK Christmas period in the hospital ward. I didn't like the holiday period in hospital. It felt eerily quiet as all of my favourite nurses, physios and doctors were away. In fact, I didn't like the weekends either for the same reason. It seemed strange to me that hospital staffing levels were so drastically reduced during holidays, while illnesses don't respect that! Never mind, signs of progress had lit up my candle of optimism and I could literally sense the beginning of the start of a possible return to normality. The strengths of my quad contraction continued to improve into the New Year. After a further few weeks of intense concentration and physiotherapy, I could finally bend my knee and lift the leg off the bed. This improvement was accompanied by an escalation of uncomfortable tingling, and a fierce itchiness sensation – indication that nerve connections and the feeling in my leg were mending gradually. It was fortunate that I experienced these hopeful signs because as lovely and nice as the doctors, nurses, physios and ancillary staff had all been to me, life confined to bed, and latterly in a wheelchair was becoming increasingly frustrating and uncomfortable due to lack of privacy and the regimented daily routine. Also, the bland English hospital food was starting to bore me. Each day, I would ask the doctor about my condition, and enquire about the discharge date. That date was unfortunately still not close, and to my further misfortune, I was soon to experience the nasty face of humanity.

A couple of girl patients had moved into the beds opposite me. One of them I recall was about one or two years older than me (aged 13 I think) who had been admitted to have her congenital shorter leg lengthened. She was bedridden as pins and plates had been put into her leg which would be gradually stretched over time. I felt sympathy for her, as the unfairness of being born disadvantaged had necessitated this painful procedure. As our eyes met for the first time, I said hello and greeted her with a friendly smile. She was not particularly communicative, which was OK since new patients are normally stressed or uncomfortable. However, it soon transpired that she was quite chatty to her neighbour while ignoring me and occasionally looking in my direction with a look of disdain. I had seen that attitude before. That's fine; we can just get on within our own bubbles. But, the unfriendly attitude was very soon replaced by sniggering with her neighbour. I could clearly hear the racist muttering between the two. It was obvious that my presence alone was offensive to her, and she was not subtle with disguising her contempt. Initially, I ignored them because it was the polite (and meek) Chinese way of the time: we Chinese immigrants just had to occasionally accept this kind of behaviour because we were new foreigners who still had a guest mentality, lacking the entitlement and righteousness of the indigenous population, or more assertive immigrants who had settled in for much longer. Another reason was not wanting to escalate the situation. Perhaps that was the wisest way anyway because when racial abuse occurs on the street, ignoring these idiots usually leads back to normality quickly as we go our own way. But we were in the same ward, opposite each other all day and night. That's it, I told her to stop it while feeling utterly frustrated by my inability to properly explain my feelings and indignation due to my limited English. Of course, as is usual with bullies, gentle reasoning never seems to work.

Her racist sniggering was just as incessant, and her insults became more direct after I dared to ask her to stop the unprovoked abuse. Her hatred was just visceral. I cannot remember if I told any adults about her abuse and bullying. Most probably I did not because I felt too proud to ask for help, and if I did, she was not dissuaded. I was getting really incensed at the ugly attitudes exhibited by these kinds of shitty people. Racism was arguably partly responsible for my current dire illness, and now even in this sanctified place for the sick and vulnerable, I was being subjected to this awful abuse and bullying. Then, during a patient visiting period while her parents (or some adult relatives) were there, her racist sniggering resumed, and her so-called adult visitors just let it happen as if it were normal and acceptable. After a week of abuse and months of being frustrated with my recovery, this was the last straw to my patience and reasonableness. For the first time in my life, after all the withholding of frustrations from my illness, and then this fucking racist bullying by another patient in front of her so-called guardian, I suddenly went ballistic. Without caring who was around me at the time, I exploded; at the top of my voice I shouted to her in the most direct of English language insults "GO FUCKOFF YOU FUCKING COW",... sobbing profusely in between my profanities ... "WHY FUCKING CALL ME NAMES? WHAT HAVE I DONE TO YOU?... FUCK YOUR MOTHER[28]..." plus some similar profanities in Cantonese thrown in between. Shocked nurses immediately came over to calm me down. I desisted my obscenities instantly, but kept on sobbing and babbling in my broken English to them on why I just exploded. There were also nurses on her side, trying to comfort the bitch because she was then sobbing from the ferocity of my rage and verbal assault.

Later, as I calmed down I felt embarrassed of my disgusting language and loss of control in front of the hospital staff and other innocent people around me; the mild and polite little Chinese kid had turned into a foul mouthed raging hothead, revealing his pent-up raw edge, and the roughness of his council estate background. As for that girl, I have absolutely no regret of my action. After that incident, she stopped her abuse. I had shut her trap. My awareness of being different, being a Chinese minority grew even stronger. I became more interested in my Chineseness, and the development of China because much of the really visceral and sniggering racist remarks stemmed from a sense of superiority; from their words and demeanour, they thought me a stupid can't-speak English foreign Chink from that backward chingchong land of China. Hence, I was often desperate to see good Chinese role models and good news about China's development. In the late Seventies, ethnic Chinese role models were rare in Western media, except for ridiculous stereotypes. News or reports about China were extremely rare too, except for the occasional footage of natural disasters. There were, however, two memorable events I watched on the fuzzy black and white TV set on my bed drawer that helped me to remember the timeline of my hospital stay: the return to power of Deng Xiaoping in the Central government in December 1978, and the short and bloody Sino-Vietnam border war during February-March of 1979. In between these singular events was UK news of an exceptionally cold winter with blizzards that blighted much of the country, and waves of industrial action which, as I recall, did not significantly affect the operation of the hospital.

After another month of intense physiotherapy, my leg had become strong enough to bear weight while wearing the detachable leg splint. The leg was still very weak as without the straight leg splint, my knee would have just buckled under my own weight. I was offered the use of crutches, but felt too proud to use them. I was finally allowed to return home in March 1979, having stayed in hospital for over 3 months.

There were other positive outcomes of my hospital stay. The enforced confinement in an English institution certainly accelerated my English learning as I was continuously pushed to understand and communicate with hospital staff and some of the patients. Months of immobility meant I had nothing better to do than to read or reflect – I particularly enjoyed Beano, Dandy and the adventures of Desperate Dan, as well as my favourite Sci-fi Chinese detective stories. My observation and interaction with a range of people in the ward provided me a greater appreciation and understanding of British culture. With the exception of the ugly bigotry of a young girl and her family, I came home impressed and grateful to the people and hospital that had nurtured me back to health.

After months of residing in a Victorian building with high ceilings, it felt strange to be back home, where space seemed to have shrunk around me, with the ceiling of our above-shop apartment seemingly within touching distance of my outstretched arms. I guess I had also grown a bit too, as laying down probably assisted my growth. The three months in hospital had felt a very long time indeed. Physically, and emotionally, it had been my most arduous period. In the beginning, I needed to see the physiotherapists three times a week to make sure my leg continued to strengthen. Initially my dad took me to the hospital which was very difficult whilst running the Chippy. Later on, the hospital was able to arrange transport for the appointments. For the next six months, ambulances would regularly ferry me and other patients to our hospital appointments. I felt somewhat embarrassed to be relying on such a public service, but at the same time I was relieved for my independence. I got to know many of the ambulance drivers and some of the regular patients who were mostly the elderly and the disabled.

The exercise routines were quite simple but tedious, and one had to have the discipline to stick with them to realise the benefits. It was quite simple really; the more regular exercises I did the faster would be my return to near normal strength. However, having H makes such progress much harder and complicated. Before I saw the physio, I had to make sure I had Factor 8 cover to ensure I didn't induce a bleed in my weakened muscles or joints. In particular, my right knee had become a highly vulnerable joint as it didn't have the strength of the quad muscles to protect it from relatively minor traumas. The problem with Factor 8 was its short lifetime in the body. With a half-life of less than 12 hours, the protection offered by it only lasts about a day or two at the most. The administration of Factor 8 is via intravenous injection, so one has to attend hospital to get the treatment. Home treatment via self-therapy was still quite rare then, due to the shortage of Factor 8 concentrates which are much easier to handle than cryoprecipitate. In any case, I thought neither my parents nor I were ready to handle home treatment. Therefore, for much of my recovering period, my right knee was often not quite right; i.e. because it had suffered minor bleeds or was in a state of puffiness. Nevertheless, after a couple of months out of hospital, I was anxious to go back to school, and to catch up with all the lost lessons. But I never liked the first day back because of the need to explain my illness, as I would inevitably tell a small white lie to avoid revealing the real reason for my absence. I was still walking with the splint on my right leg. The splint had been an indispensable aid to my getting about. It had almost felt like a natural part of me. But one day, as I casually walked past a shop-front, I saw myself walking in the reflection of the large pane of glass; it was only then I realised to my surprise that I was walking like a half-robot: my left side moving forward normally while my straightened and splinted right leg dragged its foot forward. This shouldn't really have been a surprise, but such was the assumed disposition of seeing myself as a normal able-bodied person all of my life, it had not occurred to me until that moment that visually and physically I was no longer appearing as a regular able body person anymore. From that moment, I was determined as soon as I could to rid myself of the splint. As for catching up, I needn't have worried! The school maths I had learned in HK was still a match to the lost lessons. As for the English and other lessons, my enforced confinement over the past few months had probably exposed me to more English learning than I would have had otherwise. The three months I had spent at the hospital was probably equivalent to a couple of years of intensive English language and culture tuition!

## Salt and vinegar?

For most young children, school and home life are their dominant environments which define their growing up. For many Chinese kids growing up in the UK at that time, their lives were frequently entwined with the family's takeaway business. Living above the Chippy, it was expected that everyone in the family would help out. My parents obviously took the bulk of the responsibilities in the kitchen and the counter, but as kids, there were duties which my elder brother and I were expected to help with, particularly during the busy period of weekends. The four common jobs were a) taking orders and serving the customers, b) preparing and frying chips, c) washing dishes and cleaning floors, and d) peeling and chiselling out stubborn potato skins or buds from potatoes which the peeling machine had failed to dislodge. The last job was particularly uncomfortable during the colder months when the water storing the potatoes was freezing cold while almost every potato needed individual attention to rid its stubborn buds. Most of the time I was very willing to help out in jobs c) and d) because I felt it was part of my obligation to contribute to the family, and I treated this work as exercise and building up my character and discipline. However, I only just tolerated job b) and absolutely hated job a) because these jobs were outside in the counter area, and I was then very shy and anxious not to make a fool of myself. Jobs a) and b) on the other-hand were not an issue for my elder brother, so he usually took on the shop counter duties. But when he was asleep and I had to wake him up, it caused conflict and resentment; and sometimes if I was the only helper around, I felt hopeless and embarrassed because I was too shy to serve the customers. Even when I wasn't working, living above the shop meant I could often feel and hear the bustling of customers and the fury of my dad's wok. Good for business, but I could feel my parents under pressure, and the 'heat' of the kitchen which overwhelmed me with stress too. Over time, I developed a dislike for Thursdays, Fridays and the rest of the weekends as those days represented stress, pressure, and overcoming embarrassment and shyness.

Chinese children in the UK are well known for their academic excellence irrespective of whether they attend state or public schools. This is usually attributed to their hard-work ethic, respect of education, and family expectations for good results. However, what is not appreciated is many of us also had to endure the work pressures of their parents' occupation. Growing up within the takeaway culture certainly builds up the work ethic, but it also made many Chinese kids of our generation resentful of having to share our parents' burden and losing their own freedom. This no doubt motivated many kids of takeaway families to do well academically, even though they had to work harder because of less free time. In many ways, I found writing this part of the story most difficult because it implies criticism of my parents[29], their over-emphasis of 'just-work' over a more balanced work-family schedule, of over-relying on family members and the unsociable working times associated with the Chinese takeaway culture. I was very annoyed when they decided to open every night of the week, rather than having the one evening off. Annoyed they did not employ a part-time worker over weekends to alleviate the pressure on us (and them) over the busy period; disappointed that we hardly experienced holidays or days out with them. I resented, but understood they couldn't visit me more often during my hospital stay. Of course, for me, their greatest failing of all was their failure to spot early my catastrophic hip bleed which altered my life. I do not blame them though; it was not their fault.

Of course, in the preceding paragraph I was just reflecting from an idealistic and assumed privileged viewpoint. My mum had lost much of her independence because of her lack of English, thus preventing her dealing with or understanding outside issues, while my dad was preoccupied in keeping the business running. As a self-employed business, they had to constantly compete to thrive. Naturally they treasured what they had gained, and feared losing it. Like most people busily engaged in their jobs, they didn't have the time, and to some extent the awareness to philosophise on the bigger picture. Given the circumstances of their upbringing, my parents and most Chinese parents of similar circumstances can take much credit for creating better life opportunities for them and their children, even though they probably didn't fully appreciate our stress of helping them in their takeaway while coping with academic and social issues of living in a frequently hostile school environment. And in my case, of also living with H.

## A Chinese kid in an English school

The transition from primary school to secondary school can be daunting for many children. By clever design of the early English educators, the location of the first year class of my new secondary school (Counthill Comprehensive) was just adjacent to Derker Primary, sharing the same school yard, and separate from the main upper school site. Hence the transition was not such a big a shock as all the pupils in the playground were of a similar age. Through the many new faces, it could be seen there was clearly a broader intake of students originating from a wider area of Oldham, covering the areas of Derker, through Moorside[30] to the Sholver estates. Just like before, I went through the embarrassment of having my H-condition explained to the new teachers, and then having them explain it to the whole class. I would have much preferred to have my condition hidden. By the start of the first term in the new school, my leg was finally strong enough to be free of the splint. I could almost walk normally again, although the leg was still abnormally thin and weak, whilst the skin sensation still weird and discomforting. So much so I deliberately lighted a matchstick on the skin to alleviate the intense itches, causing burn scars which I could still see today. The only advantage of having my H condition acknowledged was the excuse to skip some P.E. lessons.

The ethnic mix of the school was similar to before; virtually all the pupils were white with a handful of Afro-Caribbean and South Asian students. I was a little surprised there were not more Asian students because I had become aware of a large Asian population in some parts of Oldham. I made a few white friends and acquaintances at school but was instinctively drawn to the other minorities in the classroom. I didn't develop much rapport with the British born Black and Asian kids who were just like the other white kids with their established privilege of being culturally familiar and linguistically assimilated to their country of birth. But I remember well forming a kinship with a Bangladeshi kid called Giptansu because he was patient and had an understanding temperament. Just like before, I was the only Chinese student in the lower school, the minority of minorities. Just like before, casual racism was common by a minority of students. I had learned mostly to ignore them or respond back in a jokey or indifferent dismissive. The problem is, for some people[31] 'casual' racist insults are just not enough. There was one rather big pubescent girl in my class who wasn't happy with just calling me all the usual derogatory terms, like Chink, go back home etc. She would also pick out other ethnic features like calling me flat face, flat nose, and persistently insulting me, or anything to do with China or being Chinese. I don't believe any Chinese or China had ever hurt her or her family, and just like the girl in hospital I could not understand this kind of unprovoked contempt of another person who just happened to be of a different race. I don't know if she was racist to the other ethnic minorities, but there wasn't any doubt she had the belief that she was entitled to verbally abuse 'the chink' anytime she wished, despite my repeated warnings. I did not tell any teachers about her for two reasons: it was just not the thing to do in the cultural environment of the time; it felt wrong to grass on someone – better to resolve it between ourselves or to accept the status quo. Secondly, at the time, the teachers themselves were pretty clueless and didn't treat racist bullying as seriously as they do today. In fact, I felt some of the teachers were themselves racist.

After a term of abuse, I was preparing to end it by confronting her face on. I made sure that before this confrontation my bad leg had regained sufficient strength to allow me to move quickly should it get physical. Even though I was a boy, and she was a girl, I knew she could hurt me badly as she had a large and athletic frame and was also quite a bit taller than me. The moment was inevitable – as we queued for our lunch on a cold winter's day, on the lightly snow-covered school ground, sure as clockwork, she started racially abusing me. This time, instead of just rebuking or warning her to desist, I angrily walked up to her and demanded absolute cessation of any further insults or face the consequences. There was no point offering her alternatives unless she apologised and promised to stop her abuse. I had already planned for the worse. I would have to avoid being grabbed or being put on the ground as had happened in my previous fight in the school toilets of the same school yard. I had also learnt that English kids did not have inhibition about how they hit each other, or the ensuing consequences. However, my ethics remained intact. If she chose to confront me, and especially as she was a girl, I would have to win the confrontation without resorting to fists or any blows that would really hurt.

She did not back down and tried to grab me. Apparently she wasn't afraid, and I was sure that she would have enjoyed beating up that 'despicable flat nosed chink' as she would have seen me. Instinctively and with lightning speed, I slapped her two or three times with my palm, and simultaneously raised my left foot to her thigh to block her advance. I slapped her again twice more in quick succession as she tried again to advance towards me. Each of my slaps was carefully subdued in force in order to shock her but without causing real injury. The altercation was literally over in less than 5 seconds as she began sobbing in fear and shock at being slapped in the face. We were both sent to the headmaster straightaway. There was no sense of my wrong in my action as I angrily protested her provocations. It appeared that the headmaster sympathised with me, as he let both of us go with just a reprimand. Any physical confrontation is not pleasant, but I felt satisfied and proud that I had subdued and taught another bully a lesson. She did not seriously bother me again.

But there was a repercussion to the above event. The incident must have triggered the school to react. Without warning, my form teacher Mrs Martin came into one of my classes and announced directly to the whole classroom to treat me differently (or words to that effect) while elaborating my H condition. There was no apparent sensitivity to me being there; it was as I were invisible and dumb. Why having been wronged by racist bullies, did I have to be labelled as a special kid who should be excused from punishment because I had H. She sounded as though I were lucky and exploiting my condition. In fact, I had endured such attitude before, with the implied suggestion that I was fortunate because my H prevented them from taking out normal retaliatory action. The irony being that they were the initiator of the abuse or aggression. My only guilt to them was being different and being a Chinese kid with a special condition. I felt extremely resentful that normal healthy 'privileged' kids and adults who knew nothing about H ever thought that being a H is a privilege to exploit! Furthermore, I felt very embarrassed to be singled out in front of the whole class, and furious with the perceived hostility I encountered. The situation made me very sad and upset as the built-up frustrations of enduring H all of my life flashed by. Unable to hold back the emotional anguish, filled perhaps with a justifiable dose of self-pity, I exploded, weeping and mourning loudly and inconsolably at the injustice of life.

The above incident convinced me that my H condition was best kept secret, like I had always preferred. I didn't like to stand out or be treated differently. I clung onto the belief that I was just like any other normal healthy kid, despite the frequent pain and inconvenience. The transition to the upper school offered a new start. The school's management had learned to be more sensitive – there was no need to divulge my condition to everyone. There was already enough to endure for a Chinese boy in a rough English comprehensive school. With the exception of being excused from P.E. lessons which I didn't like anyway, the knowledge of my medical condition in school and in society had no advantage and served only to increase prejudice against me.

The upper site of Counthill comprehensive was impressively large and had the superlative of being situated on top of hills overlooking the Pennines and beautiful countryside. It had over 900 pupils who were predominately white despite Oldham having a large Asian population. The beauty of the location did not fit its often unpleasant atmosphere. The fashionable Mod style was quite common amongst the more belligerent pupils, whose domineering Red Doc-Martin boots looked both intimidating and vicious as a fighting weapon. Including my elder brother, there were only a handful of Chinese pupils; almost all of them had found themselves in fights at one time or another. In the early Eighties, one of the favourite topics of school yard talk was "Who conquers who" with the implied question that the toughest kid in the school was the conqueror. I recall how my older brother and I discussed the possibility of fighting the conqueror who was racially bullying him. As far as I know, my brother, who does not have H and was physically quite tough looking never fought him. At the time, I wished he had, as a win over the 'conqueror' would have conferred some deterrence on me too. There were other minorities at our school, but it seemed to us that the Chinese kids experienced the most stick, as the Chinese seemed to receive casual racism from the other minorities too. One reason was that we were recent immigrant kids, meaning our poor English tended to make us easy targets for mockery in the eyes of bullies, while the small number of British born Asian and Black students did not have this problem. This was not helped by the institutionalised racist media portrayal of 'Orientals' in UK and US programmes of the time, frequently portraying them in the Fu Manchu, Yellow Peril, Chinaman in a conical hat, or Bruce Li-esque stereotypes; forever foreign and inscrutable. The perception by many of the indigenous population to see Chinese immigrants as perpetual foreigners meant that many perceived the Chinese face as alien and deserving of less respect. Conversely, many Chinese felt the same, thus having a guest mentality which led them to accept passivity and even apologetic to the injustice or prejudice which they frequently encountered. Being the minority of minorities, the scattered Chinese individuals and families often had little support between them. Indeed, the sad irony was that some of the more memorable racist incidents were inter-racial. There was an older muscular black kid that would unashamedly do the ching-chong-ding-dong chinky expressions every time he spotted me or other Chinese kids. I thought him an ironic ignorant clown, who seemed oblivious to Black people's struggle against racism[32]. His persistent bigotry made me so angry that I feared that one day I would be forced to confront him, like I had done to other bullies. In case I needed to defend myself, I had been practicing Kung Fu again – punching my fists daily into the sack of dried peas in dad's shop to make them harder and tougher. Indeed, one day as I saw him behind my school locker, just before he passed by, I deliberately gave out a loud angry shriek and punched my hardest at the school locker door. It was a deliberate show of power, to demonstrate and pretend that I was really tough and capable of unpredictable violence. He seemed to get it, and just walked past without further incident! That punch was indeed most impressive as it caused a deep dent in the locker door, and my hand felt really sore afterwards. Worryingly it felt like I might have dislocated my little finger. As usual with my bad habit, I did not seek any medical attention and became rather anxious over a period of days. Fortunately there was no complication as it gradually got better on its own. Another ironic idiot clown was a small Asian[33] kid of Pakistani origin, about a couple of years younger than me, who would frequently insult me in the school playground by pulling his eyes to make the common 'Chinky' taunt. I was particularly bemused by this behaviour because he was taunting someone who was obviously bigger than him. By that time, I had adopted a more aggressive doctrine of not tolerating any crap from people who were obviously weaker than me[34]. I would respect the strong, but why should I let the weak get away with insulting me? It is not in my nature to hurt the weak or disadvantaged, but if they were stupid enough to provoke an apparently stronger person, then they deserved to be punished. The kid thought he could outrun me, but one day in the school yard I managed to sneak up behind him, grab him, cover my hands around his eyes, and PULL as hard as I could, whilst lecturing him not to ever make any further racist slurs. He cried for a long time, and I felt rather satisfied for the revenge and the lesson impacted on him. Afterwards, I did question the ethics of resorting to physical action on a younger kid, which I had never felt the need to question in my previous altercations as my opponents were older and bigger. Looking back, I had no regret because he chose to ignore my warnings, and I lacked the witty verbosity to retaliate satisfactorily, without resorting to racial abuse myself. Had I retributed with a full gamut of racial abuse back, I might have felt better with giving him a taste of his own racist medicine and making him reflect on his actions. But being an introspective person who realised the ugliness of racism and the crude hostility of racist language meant I felt frustrated by my inability to retaliate with the same degree of verbal offensiveness. It made me question whether it would have been better to rebuke racist slurs with similarly powerful derogatory racist remarks.

It is rather sad that my reflection on my secondary school experience thus far has focused so much on race and prejudice. I did not intend it that way, except the story just came out like this because those negative events in my childhood must have had a really powerful subconscious impact. But issues of race and prejudice have definitely remained a serious and lingering problem in Oldham and its State schools. There were major race riots in May 2001, and almost ten years on in 2010 Oldham's education authority became a pioneer of social engineering by trying to bring the two disharmonious communities together by merging my old school Counthill (with over 90% white pupils) with Breeze Hill school (with 95% Asian pupils of mostly Pakistani origin) on the other side of Oldham into a single school. In the new school with a new name, Waterhead Academy, lessons, sporting and extracurricular activities are shared between the two campuses, so as to bring the two communities into more regular and friendly contact. It is too early to be certain of the outcome, but I hope the experiment succeeds in bringing greater understanding and harmony. Through coincidence and events, I will return to the sensitive issues of race, culture and religion in later chapters.

By the age of 12, my childhood aspiration of becoming a doctor had dissipated. I had seen too many hospital wards and needles to really want to experience any more, even if they were not mine to endure! In any case, I had developed a pull towards physics, the science of explaining nature. I still love the nature of flora and fauna, but the lack of biodiversity in Oldham and my increasingly troublesome knee had dampened my childhood enthusiasm for wildlife exploration[35]. Instead, I started to appreciate the wonder of being able to comprehend the fundamentals of nature, the world and indeed beyond Earth into the realm of the Universe and existence. The word physics first entered my vocabulary as "Wu Li"[36] in my year 5 Chinese text books. There was an article about Einstein, with an illustration of the eccentrically wild haired genius who discovered the inner workings of the Universe. Yet he wore no socks and answered letters from little boys and girls. So genteel, kind and humble – there was an immediate fascination with this man and his enchantingly named 'Theory of relativity'! In another article, there was a piece on Chen Ning Yang and Tsung Dao Lee who were the first Chinese to receive the Nobel Prize for physics in 1957. I didn't understand the teacher's explanation of their work, except for something about left-handedness not being the same as right-handedness. It was strange that nature worked in such a subtle way, and yet how Wu Li could explain it all was just amazing! But I think the biggest influence on me for falling in love with physics came from a confluence of wonderful books and TV programmes about space; factual and fiction: 'Star Trek', 'Space 1999', Douglas Adams's classic 'Hitchhiker's Guide to the Galaxy', the sci-fi stories of Isaac Asimov and Arthur C. Clarke all made a wondrous impact on my imagination – the vast expanse of the Universe and its endless possibilities. However, it was non-fiction productions such as 'Cosmos' by Carl Sagan, and BBC's Horizon programmes about science and physics that were the most inspirational. I particularly enjoyed the human elements of discovery. In Cosmos, Carl Sagan's description of the young Einstein's route to discovering the theory of relativity seared into my consciousness; at aged 16 he imagined riding on a beam of light while observing his reflection from a mirror he was holding. He surmised that if he could travel at the speed of light, his reflection would disappear as no light from his face could reach the mirror. Further investigation of this kind of oddity, and other contradictions, led him to a series of predictions which resulted in one of twentieth century's most important physical theories about the nature of time and space. In the Horizon programme, I recall the inspirational story of Stephen Hawking; of how he faced up to a dreadful disease ravaging him in his youth and yet triumph in fighting it physically and intellectually. Science and physics especially, were really cool to me – Newton, Darwin, Faraday and Einstein were long dead scientists, and yet they were immortals whose scientific discoveries and real practical impact would continue to be relevant indefinitely. Like many young pupils inspired by science and the endeavours of great scientific figures, I aspired to making great discoveries of my own. I dreamed of having a scientific theory or a mathematical equation named after me. A Nobel prize would be fabulous too! There was no doubt that I would focus on maths and the science subjects[37] at O-level[38] in order to follow my aspirations. Fortunately I seemed to have the aptitude, as I discovered to my huge surprise that I came top in my end of year 3 (equivalent to Year 9 now) maths exam. For my O-level studies, I was really excited to be able to focus on studying all the sciences, as well as compulsory Maths and English language. But the continuation of good academic scores was by no means assured. During my O-level studying years, I had missed more lessons than most kids due to my H. Even when I was in school, I was often in discomfort and in a state of anxiety; because one or more of my joints or muscles were either recovering from previous bleeds or being at the start of a new one. The English language continued to be a challenge. This was even true for maths as new concepts had to be explained in English. Misunderstanding and misconception are more likely when there is a language barrier; over-analysing the meaning of a particular sentence often leads to further difficulties. Additionally, the lack of discipline in some classes was outrageous. I can still remember vividly the terrible bullying of a female science teacher who didn't have the skills to control the pupils. The lessons were just total mayhem and crazy to the point that I recall (with shame) even I joined in the commotion. The poor discipline shown to some weak teachers was matched by the opposite extreme of physical violence inflicted by some teachers on the pupils[39]. There were other disruptions too; my parents business had relocated to South Manchester a few months before the big exams, and in the early Eighties, the militant National Union of Teachers ('NUT') union had instigated numerous strike action by teachers, causing many lessons to be cancelled. With the benefit of hindsight, the teaching from Counthill Comprehensive School was extremely poor. In fact as I discovered later most state schools in Oldham, including Counthill were consistently near the bottom of the national education league tables. The strike action incensed me greatly as I was a keen student wishing to learn and excel in the important O-level exams. The disruptions must have damaged many pupils' chances of passing their exams, or put them at a lower grade than would have been the case otherwise. As for me, I was fortunate to be a self-motivated pupil, who really liked all the subjects, except English and French. Combined with my competitive streak to do well, I managed excellent grades A's in all the science subjects, as well as in Economics which I had taught myself. The latter result apparently annoyed the Economics teacher greatly, as none of his lower sixth formers in the year above me who took the exams that year achieved the top grade. As far as I knew, I had also been the first Chinese pupil to pass the O-level English exam in the school. However, I only managed a grade B in maths which raised doubts over my ability to follow my heroes of science, Einstein and Newton. Nevertheless, these results were the second best academic achievements of that year for the school; the best result went to my then closest friend Dominic who was awarded seven grade A's and a B. The school, however, did not seem to appreciate my effort though; on the day that academic prizes were awarded to a dozen pupils, none were offered to me. Despite this snub, there was a tinge of regret that I would not continue into the sixth form in Counthill, leaving friends and the familiar environment which I had become accustomed to.

## Hope of the sojourned lights

The excellent O-level results I attained were a fine conclusion to my time in Oldham. Despite some unpleasant times and attending a sub-standard school, I look back to Sholver estate, Counthill School and the town Oldham with mixed emotions. There are some feelings of nostalgia; this was after all my first home in a new country, where I made my first non-Chinese friends, learned to speak English, played and fished in the nearby moors and generally grew up like any other kid. But there were also times when I wished I had stayed in HK, and wished I were invincible or a Kung Fu master so I could retaliate against the then, almost daily racist abuse that was hurled at the Chinese. I was lucky to be academically inclined and motivated which enabled me to achieve well in school. But very few people really appreciated the uncomfortable and anxious life of living with H. The serious psos muscle bleed I had at 11 had left me with a severely weakened leg which made my right knee joint vulnerable to internal bleeds. In fact it developed into a so-called 'targeted' joint where frequent bleeds were common. These could be very painful in serious cases, although for most of the time, I just felt stiffness and puffiness due to the inflammation of the synovial in the joint. I had accepted this state of being as 'normal', without fully realising the damaging effect on my growing joints. I may have controlled my condition better if I had adapted to home treatment of Factor 8 which was becoming popular in the early 80's. Home treatment would have saved many long distance and time consuming trips to the hospital. Serious and even moderate escalation from small bleeds would have become rare events, as well as the pain and debilitation which accompanied them. But I refused the offer because I was too chicken to inject myself, nor did I trust or wish to rely on my parents to carry out such delicate and personal matters. I would rather leave my treatment at the hospital, in the soothing and trusting care of Sister Shaw. There would be a significant repercussion to my decision though, as the miraculous curing effect of Factor 8 treatment had a dark side which was just beginning to emerge during this time. I can still recall vividly a day in 1983 (or possibly 1982) while awaiting my treatment, I came upon an article in the H newsletter about several US haemophiliacs[40] who were showing symptoms of a mysterious disease which attacked the immune system of the body. The disease had already made headlines as it had ravaged the gay community in San Francisco. My immediate thoughts on reading this was subdued with just a bit of incomprehension as I didn't fully understood all the jargon words in the article or their clinical significance. It was obviously capable of being very serious as some of the H patients had died. But through my ignorance, I wasn't overly worried; I just thought it was yet another bloody infectious shit that haemophiliacs had to put up with from their medicines. Like the hepatitis B infection which I caught years earlier; I thought if I got it, I would be treated, and the disease would be cured. My complacency wasn't just my ignorance though; it transpired that most of the doctors and medical community at the time were also quite relaxed about the emerging news. Despite mounting evidence, many doctors on influential committees did not believe contaminated Factor 8 was the culprit of the infections. How wrong could they be – as we now know, that disease was AIDS – caused by the HIV virus which wasn't identified until early 1984. I would not know my status until late 1985 when the first UK test for HIV became available. By that time, AIDS had become one of the most feared diseases in human history. HIV went on to devastate the haemophiliac community.

Although I have never been religious, it was very tempting to find comfort in a divine being. My severe H condition often brings anxious debilitating pain that is only shared with oneself. The fear and anxiety were further exasperated in those early years, when I was very reluctant to seek help early which frequently necessitated a trip to the hospital. Despite the catastrophic lessons of the past, the 'wait and see' attitude persisted, as I often gambled on self-healing – prolonging the hurt and damage to my joints. These were often lonely and contemplative times. Often, especially when the recovery seemed glacial or non-existent, the desire to pray for well-being or the thought of the possible existence of a benevolent God who would look after me became irresistible. But my developing love for science naturally increased my resistance to the faith that there was a Supernatural Being whose existence could not be conclusively proven. The urge was frequently there, but my outlook had moved distinctly to science or at least to something which I could testify through observation. Medical science had proven its power through the development of blood clotting replacement therapy, and modern medicine had certainly already saved my life, or at least prevented the inevitable fate of becoming severely crippled, like many still do in developing countries. But even the best of modern medicine is no cure. Life would be much easier as a normal person who didn't have to endure the frequent pain and inconvenience of being a severe H. I dreamed fleetingly of miracles – wishing for God's help was out of the question now, but there was an unlikely source of inspiration from the distant hills and valleys as I gazed out of my bedroom window. At night time, I would frequently look into the sky because, as weird as it may sound, I often saw strange lights dancing or moving erratically in the distance. These phenomena were witnessed with my brothers and even frequently reported in the press. On another occasion, and perhaps most convincingly, while fishing at a nearby reservoir,

I, together with several other witnesses sighted an unusual flying object during a clear blue sky day; a highly reflective silver conical shaped flying object moving at extraordinary speed while leaving no vapour trail. I would definitely class both of these events, but particularly the second, as UFO encounters which could not be explained by any man-made invention of the time. It was popular to attribute these UFOs as crafts visiting Earth manned by aliens of more advanced civilisations[41]. I wished that were true, as I secretly harboured the crazy wish that benevolent aliens might one day abduct and cure me of H! Of course, this fantastic wish was as unlikely as me suddenly being cured by the power of Divine intervention; although it is arguable that the former has a stronger observational basis than that of the Divine Lord!

In the course of recalling my experience in the early 80's, I researched the geography of the hills and valleys that were visible from my bedroom. The distant expanse was a region called the Rossendale valley, and using the precision of Google Maps, I discovered that the sleepy market town of Todmorden, West Yorkshire, happened to be directly due north of my bedroom, exactly 10 miles away. As I have since discovered, at the time of my UFO observations, Todmorden and nearby Rossendale Valley had already been dubbed the 'UFO Alley' due to a high level of UFO activities, including a famous incident involving an alien abduction of a police constable! Hence, those sightings in my youth were definitely deserving of their mysteriousness, even though my dreamy wish of a miraculous and benevolent abduction never happened. As an aside to the anecdote, another amazing fact about Todmorden is that it has produced two science Nobel prize winners; Sir John Cockcroft (Physics) and Sir Geoffrey Wilkinson (Chemistry). Amazing!

## China! When will you wake up?

In the course of writing this book between 2011 to 2015, China became the second largest economy in the world, and in the perception of many people a new Superpower set to dominate the 21$^{st}$ century. China is now frequently in the news, whereas during my childhood I longed to hear about China or Hong Kong. That change of fortune and perception was relatively recent, as China's economy grew furiously for over three decades. This phenomenal achievement was made possible by the reforms initiated in late 1978 when the new post-Mao leader Deng Xiaoping unshackled the moribund Chinese economy. However, as I recalled my experience in the late seventies to eighties, the economic and social reforms started by Deng were raising hope, but the long term prospects of stability and prosperity were far from certain. Then, China in the eyes of my English contemporaries was another backward Third World country populated by peasants, irrelevant to their lives. Some showed curiosity by asking me about China because of their interests in martial arts, but for most, China had little influence or relevance in their lives. Their attitude was at best indifference, and at worst they saw China and the Chinese as subjects for mockery and derision. I was shocked that even in school history lessons, China was a null subject. I recall in particular that while I learned the great voyage of discoveries by Columbus, Vasco de Gama, Francis Drake etc., there was no mention of a single Chinese contribution to these endeavours. At least they could have mentioned Zheng He and great Chinese technological inventions such as the rudder and watertight compartments in ships. In mathematics and astronomy, all historical reference started from the ancient Greeks. Teachers never gave a single mention of Chinese contributions to the number systems or astronomical data crucial for subsequent discoveries. Indeed, there was hardly any acknowledgment of great fundamental discoveries by other great civilisations of the Middle East, India, and Egypt. Even some well acknowledged great Chinese inventions like the compass and printing were taught with reference to Greek[42] and Gutenberg 'firsts', rather than of Chinese[43] firsts. A common and frequent response I get is that the Chinese firsts were irrelevant, as though the West had developed all its great inventions and ideas independently, while any modern Chinese advances were inevitably copied or learned from the West. All the superlatives about power, discoveries or achievements, even in the distant past, are always about the Greeks or Romans, despite them coexisting with other great civilisations with arguably superior technology, infrastructure, governance and knowledge, i.e. from the ancient civilisations of Egypt, Arabia, Persia, Islam, India and of course especially China. The great Silk roads were simply taught in history lessons as mere trading routes for Chinese luxury goods, rather than as the world's first and greatest information highway where many great Chinese innovations were adopted by the civilisations the roads linked. Even in the neo-popular culture of marketing, one often hears claims like "French cuisine is the best in the world"[44]. Is there any wonder that ethnic minority kids are often looked down upon, demeaned and made to feel inferior or less civilised? Fortunately, even in primary school in Hong Kong I was taught about Global (Chinese and non-Chinese) history. I took for granted many great inventions and historical figures I admired were not Chinese[45], but at the same time at least I knew that China and the Chinese people had made undoubtedly massive contributions to human civilisation. I partly blame the ignorant (and racist) attitudes on a sense of superiority due in part from the manifestly Eurocentric and one-sided presentation of history in the school curriculum of the time. The school curriculum is more balanced today, but it is still very much wanting in my opinion.

I also knew that the West had been preeminent since the Renaissance with a flowering of scientific and artistic thought, heralding modern science, technology and the industrial revolution. The incredible period of scientific and technological progress, combined with aggressive colonisation, enabled Europe and later the US to dominate the world. For much of this period, even before direct confrontation by Western powers, China and indeed every other non-Western civilisation had suffered sclerosis due to complacency and bad governance. I take comfort that Chinese civilisation has remained the longest surviving civilisation of all; somehow it has persisted for over 2000 years while all other originators of great civilisations have ebbed into insignificance. Compared to northern Europe and America, the weakness and backwardness of China in the 19$^{th}$ and 20$^{th}$ century was an unexceptional cyclical trend of all civilisations. They all rise and they all fall. China was just going through the trough phase. I was just unfortunate enough to be living through this period when China's tremendous contribution to human civilisation was eclipsed by the exponential development of science, technology and commerce in Western countries. Hence, I convinced myself the arrogant and superior behaviour of many Westerners were not just grounded on Eurocentric education; they also had the advantage of chrono-centric privilege. My knowledge of the enduring character of the Chinese and her overall civilisation provided me some comfort and confidence against the frequent display of condescension from often ignorant people of all nationalities, including some ignorant Chinese themselves.

Taking pride from past glory was not very satisfactory though, because during my early years, I often asked why China, once so prosperous and advanced, had become so poor and backward? Could China even catch up with the West? Why had the Chinese, once so inventive and progressive, become so unremarkable and oppressed? Are we Chinese generally and fundamentally less creative than Western people? These were very important questions to me, my soul and who I was, not because I felt any patriotism towards China but rather a successful China would reflect better on me, an ethnic Chinese kid living in the West who, at the time, experienced casual racism almost on a daily basis. Therefore, even at aged 11, I became a keen observer of China's progress. I wished China and the Chinese people would do well because it would be good for the Chinese, the world, and for me! Ironically, I wasn't that interested in the development of Hong Kong, my birth place which gave me my first home, language[46] and nostalgic memories. I felt Hong Kong was already a world famous city with outstanding achievements and an advanced economy. I predicted that tiny HK would continue to do well. For me, it was the development of China which really mattered. If China remained as an under-achieving giant or perhaps even fell into another revolutionary trauma, then many ethnic Chinese people around the world would inevitably bear the shame and prejudice resulting from the failure of China. Liberal and right wing people alike have often suggested that ethnic minorities should just assimilate with the majority of the country, with the implication that once assimilated all prejudices disappear. The reality is, of course, not so simple; much of my bitter experience of arrogant and sometimes racist sniping is related to one's race and, even for second generation immigrants, the state of their parents' country of origin. For example, during my childhood days, when I corrected my abusers that I came from Hong Kong instead of China, I noticed many of these 'casual' racist tormentors often toned down or even ceased their abuse as though people from a more advanced place deserved more respect than if I had come from peasant China. The reality was that HK during my childhood had a much higher status than agrarian China. But it is no good saying that I was really from glitzy and glamorous Hong Kong because arrogant interlopers would claim HK's success was due to British administration. Anyway, I never had a parochial view of identifying myself as a Hong Konger, as though it was separate from China. I have always identified myself as a Hong Kong Chinese. Hence, even though I don't have much intimate attachment to the land of China, I care passionately that China will in my lifetime become more prosperous, and once again demonstrate to the world her undoubted former genius for innovation.

The above thoughts and feelings were first developed during my childhood when the common prejudice I experienced forced me to think deeply about my roots and identity. That was over three decades ago. Many great events and developments have happened since then. Arguably, the biggest world development since my childhood is the rise of China, and the shift of economic power from West to East. Despite many bumps along the way, and a serious setback I'll discuss later, my aspiration for the increasing prosperity of the Chinese people has really happened! From a backward and chaotic state in the 70s, China in little more than three decades has astonishingly become the world's largest trading nation, and once again regained the global significance that she deserves. However, China's amazing transformation over the period has not been without great sacrifice and controversy over human rights, democracy, the environment, and her development model. Some of these points are still a frequent a source of deep disagreement between the West and China. In the course of my personal journey in this book, I will confront some of these issues in the context of momentous world events that have one way or another affected every one of us. Hopefully, I will also shed some new light on the diverse thinking and prejudice between different cultures[47].

# 4. SCIENCE, RACE AND CULTURE

## The inspiration of Einstein

Anyone who has ever read about the life of Albert Einstein would without doubt find something for which they can relate. Very few people can aspire to his scientific genius, but most people would have heard of the theory of relativity, or about his contributions to the understanding and working of the Universe. His scientific discoveries were truly wide-ranging: in 1905, so called his *annus mirabilis*, he indirectly confirmed the existence of atoms through his paper on Brownian motion, conjectured the existence of light quanta (or photons[48]) in his Nobel prize winning paper on the photoelectric effect, and, of course, published the famous Special theory of Relativity which altered human's understanding of space and time. Since the time of Newton, space was perceived as unchanging and ethereal while time is an ever ceaseless flow that progresses along at a constant rate irrespective of the nature of space. Einstein showed that both space and time are not absolute entities. Space and time can contract or dilate depending on the speed of the object; a mind boggling discovery which contributed to his legendary status. But that was not all! The Special theory of Relativity only deals with uniform motion of objects moving in space and time. What about real complicated motion of bodies that are accelerating, or are subject to the influence of nearby massive bodies (i.e. gravity)? It took Einstein a further 10 years to crack this problem. It led to the General theory of Relativity which explains gravity as the warping of spacetime[49] – that planets move in orbits around their parent stars not because of a literal gravitational force pulling at them (as perceived in Newton's law), but rather that they were simply moving in the most natural paths carved by the bending of spacetime by the star. As a young teenager when I first read about Einstein[50], I could hardly understand any of his mathematical arguments. However, I can still recall very vividly the portrayal of Einstein whose intuitive and simple curious questioning enabled him to tackle the nature of gravity like no other. As I remember it, Einstein was in his office observing a window cleaner opposite; he slipped from the ladder and fell to the ground. A spark immediately lit up in Einstein's mind; when the cleaner was falling under gravity, did he really feel the force of gravity? Was there really a force that was acting on him, as though someone was pulling him down to Earth? When one thinks about it, it becomes obvious that as the cleaner was falling, he didn't actually feel any tugging forces, even though he would have been accelerating rapidly at about 9.81 metres per second every second! Wow! Such simple logic, yet it took a genius like Einstein to realise this insight and significance. Einstein then argued that if the cleaner had been falling under gravity in a windowless lift[51] which was also falling with him he wouldn't be able to distinguish if he was actually in free space, or was, in fact, accelerating toward the object (i.e. Earth) which caused him and the lift to accelerate towards it. Einstein then realised that acceleration and gravity were indistinguishable as they are equivalent. This apparently simple realisation is known as the equivalence principle which subsequently enabled Einstein to develop his new theory of gravity – the famous General theory of Relativity.

It wasn't just the science that made Einstein so fascinating to me. Even as a kid, he was a rebel against oppression and authority. I kind of identified with him as I too was a rebel but against the oppression of ill health and racism. Incredibly, at a tender young age of 17 in 1886, Einstein chooses to be stateless by renouncing his German nationality because of his dislike of how his birthplace was becoming increasingly nationalist, authoritarian and anti-Semitic. Albert Einstein once said 'If Relativity is proved right the Germans will call me a German, the Swiss will call me a Swiss citizen, and the French will call me a great scientist. If Relativity is proved wrong the French will call me a Swiss, the Swiss will call me a German and the Germans will call me a Jew.' Despite our differences, it seemed so easy to identify with Einstein, the common feeling of alienation, rejection and battles against oppression. Having developed a curiosity for science and Einstein, there was no doubt I would study physics and mathematics in my new school.

## Amazing calculus

The move to Burnage, in south Manchester was a revelation in my educational experience. On my first day in Burnage High School for Boys, in my first A-level maths lesson, I was finally introduced to Calculus, which is the branch of mathematics which enables the calculation and prediction of almost any kind of change. I was desperate to learn about it ever since first encountering those intriguing mathematical symbols found in the work of Einstein and other great physicists. I had already read the fascinating history behind this legendary topic in my early teenage years; the controversy between Newton and Leibniz about who invented it was fascinating, but I had never tried to learn it by myself because I thought it was going to be impossibly hard, and very difficult to understand. Mr Porrit, my new maths teacher introduced the concept of differentiation (a branch of calculus about finding out the instantaneous rate of change of any function) via the concept of "differentiation from first principles". For the first time in my life, I was shown the origin of an important formula and how it was derived; as opposed to just being given it. I was amazed to witness the methods of the invention of Calculus; the ingenuity, and with hindsight the simplicity of inventing such a fundamentally powerful mathematical principle. Without Calculus, modern science and engineering would have been impossible. That day and for the rest of that week, I felt dazed and awed from this learning experience. To this date, I think it was probably the most satisfying learning experience of my life. There was another revelation, the quality and clarity of the maths lessons given by Mr Porrit, Mr Hampton and Mr Samuelson were in a completely different league from the teaching in my previous school. What would have been very difficult subjects to learn before in my old school, became some of the most magical and wondrous lessons when taught by these talented teachers. The school even catered for teaching of Further Maths, even though only me and two other pupils opted to study it. I felt so privileged to have this opportunity. The physics and chemistry teachers were equally brilliant. The former was particularly memorable as he was a pipe smoking bearded Greek who resembled Aristotle. Beside his impressive knowledge, I particularly liked him as he frequently interjected his lessons with stories of ancient Chinese inventions and scientific contributions. It felt surprisingly pleasant finally to learn from a teacher who didn't just ignore non-European contributions. After a week at my new school, any tinge of regret of leaving Counthill School had evaporated. I do not have any doubt that had I stayed in my previous school, most of the wonderful learning

I had in this new school would have been non-existent. It may seem obvious now, but it wasn't until I experienced superior teaching that I realised my ability to learn and understanding difficult subjects was directly related to the quality of the teaching.

## Race and culture

There was another cultural shock in attending my new school – over 90% percent of the pupils in the sixth form class were not white. Most of them were British Pakistani, with others of Bangladeshi, Indian and Sikh origins. It was a bit of a culture shock because I had previously attended a predominately white school, and had just naively assumed indigenous white English people as the majority was the norm, as was the case in Burnage itself. Hence, it was surprising that in my new school, white English students were the minority! Initially, I felt like a foreigner again in a different continent; except virtually all the teachers were white English like before, and most of the students spoke with a distinctive Manchester-Asian accent, while I had a Cantonese Chinese-Oldham English accent. As before, the Chinese were the minority of minorities, with only about two Chinese faces in the whole school[52], and as usual I was the only Chinese student in all my classes. Did it matter that I was a minority? Unfortunately, the answer was "yes" again. To my annoyance and disdain, a small group of Asian students sitting in the opposite bench of my form taunted me constantly. At first it seemed like they were just teasing me and trying to wind me up. But their comments and actions became personal and racist: 'yellow', 'ching chong Chinaman', slitty eyed gestures, and even comments about the Chinese breeding like rabbits. WTF, I thought. I'd had so much shit from ignorant white racists before, and now I have to endure even more ignorant and infuriating brown racists. I used the phrase 'even more' because as ethnic minorities themselves, I had expected them to be more sensitive and to know better. It was difficult to verbally get my own back, because I was a minority of one and my English language skills were still weak, lacking the wit and eloquence to deflect their insults effectively. I felt that it would be much easier to insult them back with the same bluntness by using racist language and uncomfortable facts about them and their countries of origin, as they had done to me. But unlike them, I had the sensitivity and principle of not behaving like a racist, or to hurt peoples' feelings. Just like before, it never occurred for me to inform their behaviour to teachers. For most of my sixth form years, I was to endure almost daily taunts, particularly from Hussein, the chubby but strongly built leader of the group. I can't help but feel more contempt against racists who are themselves of ethnic minority background. Where was their sense of irony and sensitivity? Am I being prejudiced too for feeling more angry with non-white racists?

Beside the harassment I experienced daily within the school, there was another factor that were making me extremely frustrated and at times furious... so furious that I held fleeting thoughts of violence against the people responsible for degrading and damaging my education. No, not Hussein and his gang – that would be dealt with later – but the National Union of Teachers ('NUT') and the headmaster of the school. I still recall that ever since my 4$^{th}$ year (equivalent to year 10 now) at school, the militant NUT had instigated industrial action which led to many lessons being missed. By the time I was in Burnage in '84-'86, these waves of strikes got considerably worse. At first, it didn't matter too much because all of my teachers belonged to another Union which did not call for strike action. These teachers were all happy to continue lessons as normal. However, the headmaster of the school, who had already developed a reputation of being aloof[53] and remote, simply ordered the school to be closed. Outrageously, over a period of many weeks the school was closed up to 3 days per week! At the time, I was absolutely furious with the NUT and the insolent head teacher who demonstrated a complete lack of concern for the education of the pupils. The situation worsened even more when my favourite physics teacher decided to leave teaching at the end of the lower sixth year. The subsequent supply teacher was a nice old gentleman, but he was totally incompetent at the subject. There was a real concern that my aspiration to study physics at a good University was in serious jeopardy. Unsurprisingly, the morale in the whole of the school was just dire. Indeed, an awful event did happen during this period, the lower school's art block was burnt down in an arson attack by two lower school pupils. The leader of the attack was a delinquent white youth who had a history of trouble. He was caught, but amazingly Manchester Education authority insisted that he remained at the school, rather than be removed to a special school. I remember this event well as it was in the middle of a Further Maths lesson, when Mr Hampton, our maths teacher dramatically informed us that they would go on strike the next day to protest against the decision – the only teacher strike action which I didn't object to. There was a palpable sense the school was heading towards a crisis.

In the thick of the tension within the school, I was experiencing daily verbal harassment and racial abuse from Hussein and his friends. I had been telling them to desist for over a year, and tried to ignore them most of the time; but it just didn't get through their thick skulls. For them, it was just a bit of fun at my expense. Then, one day I had enough and confronted Hussein face-on in the classroom[54] before the teacher arrived. He did not back down, and then there was a shove from me or maybe him; I can't remember who started it, but to be honest, I wanted to teach him a lesson. So, before reaching a third shove, and without further warning and using my good leg, I kicked the fat bastard right at his fucking chubby tummy. This was a really dangerous escalation as he was a strongly built 18 year old lad. After over a year of incessant taunt, I relished the realisation that he's going to get a beating, but I'd have to win fast or I could be seriously hurt. My furious kick to his stomach did not produced the resounding effect I had anticipated – he just continued to march forward like nothing had happened while I found myself abruptly lying flat on my back. Oh bloody hell, his chubby but firm tummy had simply absorbed my kick and its rebound had literally flipped me onto the hard concrete floor of the classroom. I had underestimated his power, while not realising how weak my bad right leg had become in supporting my balance. In the midst of the adrenaline rush, while still laid on the floor, I kept my cool as Hussein relentlessly marched towards my fallen body. There was no way I was going to let him get on top of me or it would be game over for me. In the millisecond that he rushed over for his pounce, I used my good leg again kicking at his thighs and groin as hard as I could. This time, the floor provided the balance, so my kicks were probably more effective as it connected with his body more firmly. It must have hurt him as he backed off a bit. But adrenaline must have neutralised his pain and his bulky frame continued to advance on my still fallen body. I must not let him get close; I must try to stand up as quickly as possible… a myriad of defensive thoughts raced through my mind. It was at this point the fight stopped. His friends had pulled him back, while some of the other students also simultaneously surrounded me to prevent further escalation.

I got up quickly, with raging look and boiling hormones still pumping, I was ready to complete the fight but simultaneously relieved it had been interrupted. The physics teacher arrived without realising the violence which was unleashed only a few seconds earlier. The rest of the lesson was a blur as I contemplated my action and the possible consequences that could follow. The strength and balance which I thought I possessed was sorely lacking – it took a critical event of this fight to awaken me to the realisation that my right leg had over the years become insidiously weak, seriously affecting my gait and balance. Never mind, I hoped nobody would know the truth and that anyone who thought of taunting me would think twice before daring to mess with me. It was fortunate that during the fight Hussein's friends didn't gang up against me. They were after all not malicious, except for their mindless arrogance and idiotic ignorance which is so common amongst bullies and their followers.

If there was anything farcical about this incident, it was that Hussein and his friends came to my dad's Chippy the very next day, like it was just another normal lunch break. They ordered their favourite lunch grub: chips in buttered butties, and battered fish sandwiched in buttered butties for those with more pocket money. There was no apparent aggression from Hussein afterwards, although he continued to hail the occasional racist slur at me, still thinking it was funny! However, I was no longer aggrieved by his words; satisfied that he had taken two punishing kicks from me. Besides, I would be leaving the damn school soon.

On reflection, perhaps all of the racist abuse I had been subjected to over the years was just a bit of a tease. Maybe, I should have been more understanding, or even accommodating of it. I could then have justifiably copied their mockery and casually teased them back with a similar daily doses of 'humorous' racist bantering. Yep, I should have just called them back as Pakis, to those White racist White devil[55], racist Jew yid, and Afro-Caribbean bully Uncle Tom. Yep, it's all fine… grow up and don't be so sensitive or politically correct. Yeah, it's just a bit of a laugh; it's only words and words can't hurt. Just look the other way, Ding Dong Chinaman. I was probably just being too sensitive, wasn't I?

The answer to the above rhetorical question is of course an emphatic NO. Even though I was only making a point, uttering or hearing the derogatory expressions above should make every educated person very uncomfortable. Unfortunately, even today, there are still many apparently decent people who would argue that minorities who complain about 'casual' racism are being over-sensitive. Well, if you are one of them, please educate yourself about your privilege[56] instead of being ignorant. A bit of rude bantering may be fine between friends in private, but deliberate racist slurs, regardless of who makes them is simply an affront to human decency.

My school experience in Oldham and Burnage left me with a bitter taste towards the unruly English school environment with its indifference to pupils' discipline. Besides despising white and black racists in my first secondary school in Oldham, I also developed a distinctly negative perception to some of the British Asian students who were of mostly Pakistani origin. During my school and later years, I could not recall any unpleasant exchanges with Sikh, Indian, or Bangladeshi students, whereas some of my most lingering negative experiences had come from young British Pakistanis. Why? I felt, amongst the Pakistani students who had caused me grief, there was a mix of pride, arrogance and ignorance that was not that dissimilar to the white English racists that I had encountered before. I wasn't surprised by racism from the majority white population, especially as white racists can claim their unearned arrogance of perceived superiority based on the wealth, and successful achievements of their home country. For the British Pakistani racists, they appeared to me, to have derived their pride and arrogance from a sense of superiority of their own culture. I did not know or care exactly what that might be at the time, except I despise the bigotry of anyone who are racist to others. Both groups shared the common arrogance and ignorance of disrespect and condescension to people not in their respective group.

I might be able to shed some light on this from my personal experience. In Counthill School, one of the trouble-makers was Ikhlaq, a British born Pakistani kid in my form class. He, just like the kids from Burnage High had also mocked me based on my race. I tried to reason with him, and somehow we became friends. It was then that I noticed that he liked to tell everyone that he was from Saudi Arabia, instead of being a British Pakistani. He was apparently ashamed of his Pakistani origin. Perhaps he tried to claim to be Saudi because Saudis were perceived to be wealthy and powerful. I believe that a combination of low self-esteem and lack of self-reflection contributed to his misbehaviour. These negative feelings are not uncommon amongst immigrants or ethnic minorities from developing countries as they are frequently reminded of their difference from racists and the racist media, both of which were prevalent at the time. Hence, children of ethnic minorities growing up as a minority group, be they Chinese, Indians, Pakistanis, Afro-Caribbean often felt ostracised and even ashamed of their race, and their countries of origin. Self-aware ethnic Chinese and Indians can at least fall back on knowing that they can claim some comfort and pride from their rich cultures and their once great and influential civilisations – despite the regret or anguish they might feel for the relative backwardness[57] and status of the countries from which they or their parents originate. For Afro-Caribbean living in the West, they can claim much pride in their culture and achievements in the arts, as well as in politics and their preeminent role in the struggle for equality. For Pakistanis, who are the product of a country born out of sectarian and religious violence that was the partition of India, they have lost the claim to the grandeur of Indian civilisation, except perhaps clutching at their only significant identity of being Muslims. For Ikhlaq and many like him, even belonging to a proud faith may not have provided the requisite self-esteem as he and many like him lacked a common principle of respect for others, nor were there at the time, any advanced role model Muslim countries to look up to.

There was however a wind of change occurring in the mid-Eighties. By the time I attended Burnage High school, there was a noticeable increase in the influence of Islam. There was clearly an increasing pride in being Muslims as I sensed increasingly in the two years at the school. Although none of the British Pakistani student peers tried to influence me with their religion, there, in the background, was frequent talk about their religion, going to mosque meetings, prayers and things said by various Imams. Discussion and mention of words like Sharia, Halal, Haram, Hadith, Jihads, and Kaffirs etc. were commonly heard. There was a growing sense of pride of being a Muslim because of events that were shaping the world at the time. In fact, before I had shown any indication of interest in Islam, I recall strongly my admiration and respect for the ragged and brave Mujahedin fighters who were at the vanguard of the Cold War. Against all odds, they fought and defeated the might of the Soviet army in Afghanistan. Unlike today, these Muslim fighters which include Osama bin Laden were allies of the West. Pakistan played a pivotal role in supplying weapons and fighters in the common goal of defeating the evil[58] Soviet empire. So how is this relevant to my experience? I believe brands of fundamentalist Islam such as the Deobandi movement and Wahhabism sponsored by Saudi Arabia[59] and indirectly by its Western allies had led to an increasingly radical Pakistan. I think this sowed the seed of Islamic radicalisation in Pakistan. Subsequently, this malign influence unsurprisingly seeped back into the community in the UK. I sensed this as far back as the early 80's during my school years – not particularly in the way of religious fanaticism yet, but I had a strong weariness for their arrogant and exclusive attitude. Much later, I learned that devout Muslims unquestionably believe that Islam represents the absolute truth and that Muhammad is the last messenger and supreme prophet of God. Could this be the source of pride and arrogance which led to the racism I experienced?

I have always thought that as ethnic minorities, we should make the best of ourselves by enjoying and combining the best of British with that of our other cultures. Ethnic minorities are just like any group of people; some choose the best of both cultures, some choose the opposite, but most are somewhere in between. It is important to realise that the best and worst of anyones' cultures are not mutually exclusive. Virtue is an individual choice derived from education and inner compassion. No single group of people or beliefs hold the monopoly of virtue. In the impressionable minds of young people where religion becomes a significant part of their culture, the proud certainty and assertiveness in their beliefs are often manifested in contempt towards non-believers or people of another denomination[60]. Combined with the ingrained, and often poorly educated traditions of their parents' homeland, I felt many British born Muslims growing up in the 80's, and subsequently were conflicted. Hence, as a community they had some of the worst social and economic problems among ethnic minorities. The situation is arguably worse now, with jihadi inspired radicalisation being major problems within a significant minority of the British Muslim community. I hope my Pakistani friends, and acquaintances do not take offence by my frankness[61]. The reality is all cultures have faults and the sad fact is much of the Muslim world is under siege from appalling extremist indoctrination. I feel strongly that liberal and progressive Muslims must actively counter the malign and extreme influence of radical and political Islam (i.e. Islamist ideology) or all of us will suffer further from an inevitable increase in violence and prejudice in our world.

## Triumph and tragedy

The summer of 1986 was an exciting time, with great anticipation of going to University. Despite the usual H related obstacles and all of the school's disruptions, my love of physics and maths continued to motivate me as a successful student. Diligence and hard work meant I had easily surpassed the A-level grades required to study theoretical physics at the University of York. I obtained the school's best A-level grades, beating a rather aloof and snooty classmate who was previously the school's top academic achiever. He looks rather like a young Stephen Hawking with his NHS style glasses, middle class accent and thick pair of lips that is rare for the average white kid, but for some reasons quite common in white people with high intelligence and posh background[62]. Through jealousy or envy, he would sometimes accuse me of being a swot, even though he knew nothing about my other interests or indeed the extra hurdles I faced daily with my health and the responsibility of helping out in my parents' takeaway. I've told this anecdote to highlight a particularly common attitude of some British white students.

That is they like to pretend or be perceived as if they have achieved their usually mediocre academic results effortlessly. Often, these students would imply or claim that they could have done better had they worked harder instead of enjoying their numerous extracurricular activities or messing around with mates. There is in particular in Western popular culture an association of cool with effortlessness. This attitude is sometimes extended to debates concerning the performance of Western vs. Chinese students in general, with the implication that Western students are actually smarter, if only they worked as hard. Well, whether this is true or not is irrelevant because in my experience, the best of Western students or employees work at least as hard (and often harder with more focus) than the best of the Chinese. As Thomas Edison once said "Genius is 1% inspiration, 99% perspiration". Anyone can fail or be mediocre because they don't work hard enough.

Working hard was certainly not in the mind of one particular pupil. On September 17th, 1986, one of the most notorious events in the history of school violence occurred in the playground of Burnage High school. The kid who burnt down the school's art block a year earlier had pulled out a knife in a premeditated fight with a Bangladeshi Asian student. He stabbed him, and the Asian student bled to death in the playground. An innocent 13 year old boy had died by the hand of a violent and badly disturbed boy. The assailant was White, and the victim Asian. The tragedy became headline news for its exceptional violence inside the school, and was further exasperated by the racial element. When I heard the news, I was shocked and saddened but not entirely surprised. The morale and ethos of the school had been falling steeply even in the brief two years that I was there. As I have described in my own experience, the culture of name calling, bullying and indifference by the school's senior staff was endemic. The MacDonald inquiry[63] revealed that racism contributed to the murder of the pupil though the whole story is complex. In my opinion race was incidental to this brutal murder by a dysfunctional and disturbed child. The tragedy could have been avoided had Manchester education authority not refused to remove the boy from the school after the previous year's notorious arson fire. The top management of the school was also blamed for their weak and indifferent leadership of the school. I couldn't agree more. As I made the point earlier, I felt despair and absolute rage when the headmaster ordered the entire school to close (for up to 3 days in a week) during strike action called by the NUT, even though none of my subject teachers belonged to that union. Those prolonged days of school closures and disruption in the 80's must have adversely contributed to the appalling moral deterioration and tension within the school.

It is ironic that as I write this part of the book in 2013, teacher Unions are calling for industrial action. I would like to plead with all teachers contemplating industrial action not to take any classroom strike which directly affects pupils' learning. Boycott marking or some extracurricular activities if you have to, but please do not close down classroom lessons as it will do untold harm to many pupils.

## The British Chinese

Having made some critiques of other peoples' cultures, perhaps this is a good time to review the cultures of the British Chinese community. In the Eighties, most of the British Chinese population were immigrants from Hong Kong with many originating from the rural areas of The New Territories bordering China. Most were Cantonese speakers, although many of the latter group were 'Hakka' people speaking a different dialect. Unlike some ethnic groups, the British

Chinese don't tend to form segregated communities and are geographically dispersed throughout the country. Chinatowns, though famous around the world are predominately business centres with restaurants and shops rather than a place where Chinese live. This speaks much about the British Chinese characters of the time – they were perceived and tended to be self-sufficient. This is a positive point in my opinion, but many British Chinese were also isolated and insular in the sense that many were busily engaged in the catering trade, leaving them with little time and motivation to engage more broadly with British society. The focus on work is a strength for economic survival but suffocating for the younger generations who are often expected to help out in the business. On top of this responsibility, we were also expected to do well academically. This duel pressure was often a source of conflict between siblings as they argue about who should do what, and between parents and their kids during the high pressures of a busy customer queue, or in the heat of the kitchen when tempers fray. Thus, many Chinese kids were motivated to do well in school in order to escape the lifestyle of their parents, but it's probably also true that many also suffered academically because of their family responsibilities. Being dispersed and usually non-confrontational, the Chinese have arguably been subjected to, and tolerated much more casual racism than any other ethnic group. This high tolerance for abuse, and the propensity to not complain about it, is for me a particularly weak point of the Chinese character. Another particular difficulty faced by many British Chinese is the language barrier. Obviously there are little outstanding language issues with British Born Chinese, or those who arrived in the UK as young children, but for the older generation immigrants, or even younger ones who work exclusively in the catering trade, becoming competent in the English language is an extremely hard goal for them. Ironically, advances in communication and entertainment technologies have further hindered immigrants' progress in learning the language. I remember the widespread adoption of the video cassette player in the early Eighties dramatically reducing my whole family's uptake of English TV programmes as we eagerly lapped up Hong Kong produced dramas. Subsequently, there followed satellite TV, DVDs and now internet TV. As people always desire familiarity, a significant consumption of material in immigrants' original home-language is inevitable. Thus, technology has to some degree hindered integration, although, in the end, I think technology will also in time enhance communication too.

On the subject of culture, although many Chinese kids of my generation have thrived in the professions, unlike Black Britons, the non-cohesive British Chinese community lacks a strong voice or narrative to develop any significant distinct British Chinese sub-culture. Thus, the British Chinese has not significantly contributed to the arts and cultural scene of Britain. Part of the reason is the lack of critical mass in the numbers of Chinese wanting to be engaged in the arts. The other aspect is probably cultural as there is too much emphasis within Chinese culture to making money as the primary marker for success. Having said that, Chinese immigrants generally do well in most countries they make their home while their children tend to succeed and integrate without difficulties.

# 5. FREEDOM AND SUFFERANCE

## In the shadow of disasters

The widespread introduction of effective treatment in the Seventies for the terrible suffering caused by H was an enormous relief for H's living in industrialised countries. Receiving the pale clear Factor 8 liquid into my vein almost always stabilised the internal bleeding in joints or muscles. When I was young, its effect was almost immediate and bestowed an aura of miraculous-like recovery. The subsequent widespread affliction of hepatitis B on the H community only dampened slightly the enthusiastic embrace of Factor 8 as the gold star treatment. But by the early Eighties, a handful of H's started to develop the deadly symptoms which were similar to the ones that were devastating the San Francisco gay community. It was AIDS. During those early years, there was very little information from the medical community, no definitive tests for the disease and little precautions given about it. Despite evidence that AIDS could be transmitted through blood and blood products; there was complacency that H's were unlikely to catch AIDS through treatment. It was comforting to accept this belief – after all, doctors know best. In any case, with the exception of the accustomed pain and debilitating effects of H, I felt immortal as most healthy teenagers do – I was sure that AIDS was just another blood borne disease like hepatitis which even if I did catch it, I would be able to fight it off. That view persisted even when AIDS became headline news. The revelation that Hollywood legend Rock Hudson had the disease suddenly propelled the illness to global awareness and notoriety – then often portrayed pejoratively as the gay plague that could rapidly turn a healthy young man into an emaciated wreck. Everyone who showed the symptoms died from it. The number of H's showing symptoms was still very small, but the association between AIDS and H was becoming more common by the day. The case of Ryan White, the Indiana teenager and H who had contracted AIDS became international news as he fought prejudice and stigmatisation when he was refused entry to his school. Even though I remained relatively fearless of the disease, I became even more uncomfortable that H was now further stigmatised with probably the most feared infectious disease of the modern world. On 2$^{nd}$ October 1985, Rock Hudson died of Aids, and I was about to find out the result of my first HIV test – the test having just become available in the preceding month.

Amidst the alarming reports of Aids victims around the world, the waiting for my own results was surprisingly carefree. I hadn't got a family of my own, or been in an intimate relationship with anyone, so there was no worry for anyone else except myself. I had no intention to tell my parents of the test as they didn't yet know the true extent of the AIDS crisis with relation to H. I did not have many sleepless nights over the imminent results. The years of frequent nightly worrying over the direct pain from H had internalised in me a philosophy of 'peace and acceptance'. Whatever my fate was I would just have to face it. Anyhow, I remained unreasonably optimistic and blissfully ignorant at the same time; not quite realising that if I were tested positive, the ramifications to my life would have been unimaginable. I entered the familiar H treatment room at the Royal Manchester Children's Hospital (RMCH) – doctor and nurses were there as normal. I think it was Dr Evans who broke the result to me. With a reassuring smile, I was told the good news- I was HIV negative.

I was relieved, but not overjoyed as it wasn't like something good had happened to me, but rather that it felt like I had missed a train which then crashed. In due course, I realised I was extremely lucky as 1,246 H's in the UK alone were diagnosed HIV positive in 1985. Over the coming months and years, more and more H-patients fell seriously ill. Of these, over 800 have since died. Before these tests, few could have imagined this impending human tragedy. Many of the victims were children and teenagers, some were moderate H who rarely needed treatment, but were infected with just a single bad batch. The anger and despair of many H's and their families was completely understandable and palpable. Many H's became activists.

As the death toll mounted, I thought about why I had been so fortunate to have escaped this calamity. The coincidence that RMCH continued to use UK sourced plasma Cryo-precipitate (instead of imported Factor 8 concentrate from the US) due to its cheaper cost had ironically prevented many of its patients from being infected. My stubborn refusal to accept regular prophylactic treatment (which would have used Factor 8 concentrates), or even urgent treatment for some bleeds had led to much unnecessary damage and painful episodes to my joints and tissues, but may have ironically saved my life as I had minimised my exposure to infectious batches. With hindsight, perhaps all the miserable pain I had endured for denying my own treatment may have ultimately saved my life.

What about the future? Could I be infected by a future bad batch? That possibility has been much reduced since 1985 as all blood donors are now screened, and furthermore Factor 8 is also heat treated to deactivate most known pathogens and viruses. However, for as long as Factor 8 is still made from human sources, there was always at the back of my mind a lingering anxiety that I could at any time be infected. Or indeed, could there be other, as yet unknown dangers lurking in blood resources? This anxiety wouldn't be extinguished fully until genetically engineered synthetic Factor 8 became available almost two decades later, in 2004.

## Physics and independence

My passion for physics enabled me to go to the University of York to study theoretical physics. It was an ideal subject for someone who didn't like experiments, and if I were successful at it, an ideal career path for an H. Leaving home the first time for most teenagers can be daunting; losing the comfort of the family home and starting the journey of independent living. I didn't live in a traditional family home though – rather a home on top of a shop, so the feeling of stress was never far away. I was glad to leave that behind. Choosing York was also a strategic decision as I picked it not just for the subject, but also because it has a lovely campus in a beautiful city. It is also not that far from Manchester in case I fell ill. Since I had not learned to administer Factor 8 by myself, there was additional trepidation and worry of how to deal with my H condition. I would have to go to hospital every time I needed treatment. This hassle was a frequent excuse to defer treatment. In fact, I recall on my first night at University my problematic right knee 'puffed' up again. It had become inflamed and wobbly due to recurring bleeds inside the joint tissue. Ideally, I should have had it treated and rested, but it was my first evening at University and I felt I needed to fit in by joining the pub crawl into and around the city. So I just put on my trusted tubi-grip (a type of compression support bandage invented in Oldham) on my knee and joined in. I put on a friendly act, making casual introductions with a bunch of confident and predominately middle class white students. But it was not much fun as I trod along with these carefree and excited freshers, whilst actually worrying if I was going to be totally debilitated the next day – and I don't mean from the alcohol! I left the scene early. It was a relief to return to my dorm room, immediately resting and applying another layer of bandages around the knee to ensure a comfortable compression while I rested and tried to get to sleep.

The next morning, I woke up relieved that my knee had stabilised. That wobbly and mildly inflamed knee had, in fact, become part of my 'normal' state for a few years already. The sensitive tissue around the knee joint had made it a 'targeted' joint meaning it bled easily. The only way to prevent it was to have regular doses of Factor 8 in my blood stream. The treatment required injecting Factor 8 three or more times a week to ensure there was always sufficient amount of the factor in the blood stream, thus preventing spontaneous or aggravated bleeding within the joint. Given my still rebellious attitude to treatment, that wasn't going to happen. I did not fancy going to hospital three times a week nor did I wish to learn to inject myself intravenously. This resistance was allowed to persist for most of my undergraduate years because my semi-permanent puffy right knee was relatively free of acute pain. Most of the time, University life was great despite this 'inconvenience' which I had just bloody mindedly ignored; a prolonged mistake which I would come to regret deeply.

With Einstein being my childhood hero, studying physics at University was really a dream come true. There were many memorable lectures in physics and mathematics covering topics which I had only previously read about (but mostly didn't understand) and now I was actually going to be taught them from University professors. In physics, the syllabus consisted of exciting subjects like atomic physics, special relativity, general relativity, quantum physics, thermodynamics, kinetic theory, and electromagnetism, which I had so much wanted to know and understand. Similarly, in mathematics, I got to learn about Fourier analysis, delta functions, partial differential equations, and Laplace transforms. Physics and mathematics hold such fascination to me because they explain so much about the fundamental nature of the world. Compared to art subjects, science is more absolute as one tries to uncover the secrets of nature, whereas arts are subjective and arbitrary. Funnily enough, despite all the exciting fields studied, it was the very first University physics lecture on the first day of teaching on a relatively mundane topic of 'dimensional analysis' taught by Dr Jim Matthew which has somehow most impinged on my memory. How? Like most aspiring theorists, I fancied coming up with new equations of my own. In one memorable example, while still doing my A-levels, I proposed an experiment in a competition run by the BBC's Tomorrow's World programme, to test whether there is any validity in psychokinesis. My idea involves measuring the time of swings of a carefully set-up pendulum in a crowded studio, where the audience is asked to mentally trying to slow down the pendulum. My analysis led to a complicated and rather cool equation which predicts how the time (or period) of the swing could be related to the 'apparent' force transmitted by the audience. Well, armed with the simple principle of dimensional analysis, I recheck the equation which I had previously deduced with great pride and logic; as it turned out, the equation was total nonsense. On re-inspection, my derivation must have been erroneous as the equation for the period (i.e. 'time' variable) did not have the 'dimension' of time! Basically, in dimensional analysis, an equation representing any particular physical quantity must have the correct corresponding dimension. For example, suppose I have deduced an equation about distance. Since the dimension of distance is [length], then the equation which could be made up of various variables (e.g. speed, time, energy, force etc.) on the right hand side of the equation must combine in such a way that it must also has the dimension of distance i.e. [length]. If it didn't, then that equation is nonsense. Similarly, an equation about time must give a dimension corresponding to [time]. When this simple principle was understood, I felt a great sense of revelation as well as feeling rather stupid that I didn't know that before. Prior to learning this, what I thought was reasonable argument was in fact just non-rigorous and muddled physics. There were to be many similar revelations during the course of my higher learning.

I made many good friends at University. One of my closest friends in the beginning was another Oldhamer, Chris. We connected because of our shared love of physics as well as our Northern roots, sharing much of the humour and joy of discovery. He went to a private school though, although he was from a modest middle class background because he mentioned how badly he had been fed previously because his parents had spent so much on his education. I suppose it was a very wise decision, as Oldham's state schools had some of the most abysmal exam results in the country[64]. He went to Oldham's best academic achieving school Hulme Grammar, where he must have been a contemporary of TV's most famous physicist/broadcaster Professor Brian Cox! Another of my best friends from the physics class was a South Korean called Don, a 1.5G immigrant[65] like me. He was also educated from a private school, and unusually managed to speak fluent

English without any hint of his root. Up to that time, I had not known anyone who went to private school – except often hearing the anecdotes of Manchester Grammar pupils being harassed in bus journeys by kids of Burnage High. There was a sense of self-congratulation when I realised that I shared the same class as them now – without their privileged and expensive education which got them there. But the reality was slightly different of course – they all possessed a certain know-it-all confidence and experience about the world which is often lacking in pupils from a comprehensive school background. According to some statistics I have seen, even though we are now on the same footing academically, their advantage in their career achievement on average will persists over their lifetime. Furthermore, the disparity is really much worse as hundreds of thousands of comprehensive pupils never got into University in the first place.

One particularly great thing about University life was for the first time in my life amongst young people, casual racism was virtually non-existent during the whole of my undergraduate study[66]. Ironically, my best mate Don was the one that constantly hacked on about how great Koreans were compared to the Chinese! It was a bit of bantering between mates, so I kind of tolerated it as I could give back as much as he gave – a common defensive mentality of people from small nations! We communicated in English of course, but when we were sharing a meal together, eating rice with chopsticks and despite withstanding the pungent smell of Kimchi, I was awaken to how much I have missed out previously of not having had many friends of similar cultural background.

In my first year there weren't that many Chinese students at the time, so every East Asian student that saw each other tended to be extra friendly. Whenever we passed each other, we would wave and acknowledge each other. Thus, I made friends with a small handful of Hong Kong students. Simple pleasures like sharing s meal together, talking with them in Cantonese and enjoying our freshly cooked Chinese food were common activities which were particularly standout for me. For the first time in almost a decade, I was reconnecting with the people and culture of my birthplace. There was one particular conversation which I recall strongly – it was about China, how it was finally moving in the right direction under the economic reforms of Deng Xiaoping. I told Po, who was a 28 year old HK civil servant studying a Master's degree in Social policy, that relatives in my parents' home town were getting wealthier all the time. Many of them had recently bought Black and White televisions. But Po said to me, despite this progress, such was the size and depth of poverty in China; he reckoned that few Chinese would even have flushing toilets in two or three decades' time. I was really disappointed to hear that as two decades sounded like an eternity to a 19 year old. Po's prediction was considered realistic then, but his prediction has been proven too pessimistic, such has been the pace of economic growth in China since then.

    University is of course also about meeting the opposite sex and maybe the love of your life. There were certainly many pretty girls I fancied, and for some reason they happened to be Americans. One was an American Vietnamese girl and the other an American Jewish girl (I know because she had the Torah on the windows of her first floor student room). I won't go into details except to say shyness, and conservative mind-sets played some part in my failure. The latter point is of particular interest in the sense that whilst I was not religious, at that time I upheld a very conservative and traditional view that dating was of utmost seriousness which should only be conducted with a potential long term marriage partners. Sex before marriage would also be wrong and immoral. I suppose these were good virtues, but over time, I realised my mind-set on dating and sexuality was stuck in 1970's conservative Hong Kong Chinese morale, whilst younger and even contemporary Hong Kong and mainland Chinese young people I met had seemingly progressed beyond their conservative mores. That aside, the biggest stumbling block was my H condition – I was then (and even now) still extremely coy about revealing my condition to anyone; neither Chris, Don nor any of my York friends knew about my condition. Even to these male friends, I hated that I frequently had to lie to them as to why I was limping, or why I sometimes had to shut myself in my room instead of socialising with them more often. How was I ever to avoid the inevitability that one day I would have to divulge the unspeakable to a potential girlfriend? Perhaps, it would have been easier if a girl chase me instead – then I could choose the right time to divulge my condition. And even then, I think it would still be incredibly hard to do. For a more mature and open person, having H and being open about it should probably not be such a big deal. However, the onset of the HIV epidemic had cast an extra stigma to having H. I was fortunate being HIV negative, and free of the other potentially serious Hepatitis B virus which I managed to clear in my childhood. But I also knew that I had been diagnosed with non-A, non-B hepatitis, a generic term assigned to people with abnormal liver function. It was not considered serious at the time[67], but this nevertheless also caused a block to my relationship development as I feared transmitting something bad to a potential partner. For these reasons, despite some opportunities, I never started a relationship during my University years[68]. Don't feel sorry for me though; I was a rather independent person who quite liked solitude. In any case, there was consolation in my appreciation of physics and current affairs; two areas which make life so interesting.

## 'God does not play dice'

Einstein once famously said the above quote about quantum mechanics – the most successful physical theory in physics that has proven itself in numerous experiments and applications. Although Einstein himself was a founding father of quantum theory[69], he grew increasingly disillusioned with the theory of quantum mechanics because physical properties (e.g. position and momentum of a particle) according to quantum mechanics can only be predicted via a probability interpretation. Furthermore, according to Heisenberg's uncertainty principle, an object cannot simultaneously have certain properties, e.g. it is impossible to know both the position and momentum of a particle exactly. For Einstein, the world of quantum theory was simply not right. It led him to develop several ingenious thought experiments to disprove the validity of quantum mechanics. Most famously his titanic struggles with Niels Bohr, another pioneer of quantum theory, were well documented. Every time Einstein proposed an ingenious experiment attacking the integrity of quantum mechanics; every time he eventually lost the argument to Bohr, including one time when Bohr applied Einstein's own General Relativity theory to defeat Einstein's argument. But Einstein did not give up easily.

In the spring term of 1989 I was very fortunate for my final year project to have the opportunity to do my first ever physics research project about Bell's theorem which delved into the nature of quantum mechanics. Bell's theorem was inspired from the famous 1935 scientific paper by Einstein, Podolsky and Rosen (or the 'EPR' paper for short) entitled "Can Quantum Mechanical Description of Physical Reality Be Considered Complete?" Having failed to prove the inconsistency of quantum mechanics, Einstein and colleagues were able to formulate a problem which apparently demonstrated a negative answer to the above question. This time, even the great Bohr was struck; "This (the EPR paper) onslaught came down upon us as a bolt from the blue" remarked Rosenfield, a colleague of Bohr. However, Bohr was eventually able to refute EPR's argument using arguments based on a quasi-philosophical interpretation of quantum mechanics – the so-called Copenhagen interpretation which roughly says that it is impossible to ignore the role of the observer and the measuring apparatus when uncovering the nature of any (elementary) object. Such was the success of quantum theory and the authority of Bohr, Einstein's concern was pretty much marginalised and ignored until 1951 when the arguments of EPR can be appreciated more readily from a thought experiment proposed by David Bohm. He considered a pair of spatially separated particles originally prepared in the so called spin singlet state, i.e. before the particles were separated they were in a combined system with zero total spin. According to quantum mechanics, each of these particles when separated possesses a quantum mechanical property called spin, where the spin components of each of these particles can be measured independently along any direction. Hence, if one measures the spin component of either particle, then one can predict the spin state of the other particle *without* disturbing it, e.g. if particle 1 was measured to have spin-up (say) then without measuring particle 2 we can predict with certainty that it will have an opposite spin (or spin down) because quantum theory predicts it. Quantum mechanics cannot predict if particle 1 is in the spin up or down state before the measurement, but once the measurement is made and the experimenter knows its spin state, then as if by magic, another experimenter with the same setting of the measuring apparatus (i.e. measuring the spin about the same axis) would always measure the opposite spin result for particle 2. This 'entanglement' property of the two particles is a fundamental nature of quantum mechanics, but it is extremely strange that such entanglement should persist or exist at all once the particles have separated- indeed the particles could be many thousands of miles apart when such measurements are made. How can one particle be instantly correlated to another at all times and distances?

Consider the last question above; according to quantum mechanics, whatever spin component is first measured in particle 1, the measurement of particle 2 always produces the opposite spin when measured along the same axis. Could there be some form of communication between the particles, or as Einstein put it: some 'spooky action at a distance' happening? Another view to consider is that quantum mechanics is incomplete. That is, these particles were actually created with definite opposite spin about every axis. Hence, there follows the inference there are some *hidden variables* associated with the properties of the particles that are missing in the current formulation of quantum theory. We can examine these problems another way by considering the following; what if the experimenter who was making the measurement for particle 2 decided to measure its spin component along an axis different to particle 1? According to quantum mechanics, particle 2's spin state is completely random, with a 50% chance of spin up or spin down. But how does particle 2 know that particle 1's spin state was measured along a different axis for it to give rise to a random result, and also to give a definite opposite result had particle 1 been measured along the same axis. Hence there must be some form of action at a distance (or 'non-local' action) between the particles, or that particle 2 knows more information than quantum mechanics allows via additional hidden variables.

Subsequently, a number of hidden variable theories were formulated, although none of them were very compelling. However, world attention to the 'EPR paradox' returned with vengeance in 1964 when Northern Irish theoretical physicist John Bell showed in a series of brilliant papers that *no hidden variable of any kind can reproduce all the experimental predictions of quantum mechanics*. This is the famous Bell's theorem which was very significant because it led the way to practical experiments that would verify the nature of quantum theory. Subsequently these experiments have been shown to agree *completely* with quantum mechanics while part of the predictions made by the hidden variable theories differs from experimental observations.

What does this mean? As the predictions of quantum mechanics is correct then we must accept that 1) somehow 'spooky' action at a distance (non-locality) is happening in the quantum world, and/or that 2) the concept of physical reality as commonly perceived[70] doesn't really hold. Personally, I don't like 'action at a distance' as a plausible explanation. Could one really believe that two elementary particles separated by any distance could communicate and then conspire to act with each other? Hence, abandoning conventional understanding of realism[71] is the option that I have chosen – which is tantamount to seriously questioning some apparently strongly held views of the world, such as "Is the moon really there, when nobody looks?"

With much gratitude to my supervisor Dr Tony Sudbery, I felt extremely privileged to have been given a taste of the excitement and challenges of this area of research. I hope some readers have also felt that sense of wonder too[72]. Do you agree that one needs to abandon realism? Or perhaps you wish to remain highly sceptical like Einstein – who once wrote "Quantum mechanics is most awe-inspiring. But an inner voice tells me that this is not the real thing after all".

My project also investigated Bell's theorem when extended to a three particle system, which actually anticipated some later famous work of GHZ[73] entanglement. Unfortunately, this was only a one term undergraduate project so my investigation ceased at Easter as I prepared for the final exams in the extraordinary summer of 1989.

## China; hope and despair

The summer term of 1989 was a very important time personally as the culmination of three years of undergraduate study was coming to a conclusion. The extremely competitive nature of theoretical physics meant getting a first class degree was a must in order to gain entry to further graduate studies. Unlike earlier times when I was confident and had rather enjoyed the competitive nature of exams, the preparation for the upcoming finals felt extremely torturous and stressful. Up to the time of the finals, I had done well and was expecting to achieve a first class degree. However, the finals were worth more than 50% of the overall score and for the first time in my academic life I felt the overwhelming pressure that in order to achieve the goal, I would still need to do consistently well throughout all the exams. I became really nervous and panicky that I didn't have the intellectual reserve or enough time to revise effectively – these exams were particularly hard because each paper covered several subjects at once, and some of the exams occurred on consecutive days. One had to revise so much in such a short time. I felt in no doubt that all students could have achieved much better marks if only the University had spread the exams more evenly to enable more focused revision.

Amidst this crisis of personal confidence, I was also enthralled and distracted by the momentous events happening in China. The sudden death in mid-April of Hu

Yaobang, a deposed Communist party secretary who was a liberal reformer and former protégé of Deng Xiaoping had brought an outpouring of grief and demonstration by students in the major cities and campuses in China. The rapid economic development of China in the past decade had transformed Chinese society greatly, but inflation, limited opportunities and corruption by officials exasperated the grief of the people to protest for more and faster reform. More protests and demonstrations started to spread through to Shanghai and other provincial cities all over China. Most spectacularly, Tiananmen Square was occupied by protesters initially led by students but later also joined by workers and other ordinary people of Beijing. To my surprise and pride these protests were tolerated week after week through April and May. In the square, there was singing, joyous debates and a sense of achievement that the demonstrators were melding the authority to listen and negotiate. The situation was very tense though, as not all the demonstrations were peaceful. Nevertheless, for an intense few weeks the authority appeared to be tolerant of dissent. Previously reclusive leaders were even forced to negotiate with student leaders who famously appeared in their pyjamas, while pro-reform leaders such as Zhao Ziyang even appeared to side with the students in an emotional speech amongst the students heeding them to diffuse their protest. There was excitement and hope that under moderate party leader Zhao Ziyang, China would not just be the pioneer of economic reform, but may well soon be a pioneer of social and political reform in the Communist world. But it was also clear that hardliners such as Prime Minister Li Peng were having none of it. The continuous protests caused major disruption to much of China, and there was genuine fear that central power may even disintegrate leading to civil chaos or even civil war.

As an outsider but passionate observer of China and her progress, my initial support and pride for the protesting students and their cause became wearisome as it was becoming obvious they (some of the student leaders) had become factious and arrogant for their refusal to heed advice to back off. In my opinion, they had won the battle for sympathy against the intransigent rulers in their initial protests in April and May, but their feckless desire to continue the battle was a grave mistake. By mid-May, I felt the students should have ceased the protests to enable Zhao Ziyang's liberal reforming faction a chance to win over the hardliners such as Li Peng. At this point, I believe the paramount leader Deng was still sitting on the fence, although his instinct would be to ensure that China didn't once again fall into chaos. The escalation of protests by students during Gorbachev's historic visit to China not only humiliated Deng, but strengthened hardliners such as Li Peng to view the protests as a counter revolution. There wasn't any doubt by then that Deng had firmly aligned with this view – that China was on the verge of a counter revolution and chaos. While I do not have much sympathy with communist rhetoric or talk of counter-revolution, many ordinary Chinese people agreed with the prognosis that China was heading to chaos if order was not re-established soon. On 20th May, Deng gave the order – China was under martial law.

Beijing was surrounded by thousands of soldiers and tanks – the romance of idealistic youth with their spirit of freedom now faced the gun barrels of armed soldiers and their armoured vehicles. As soldiers entered the city, tearful students and workers rushed to them, pleading them to return and hold back. Amazingly, entire infantries of soldiers held back, and some of them even retreated. It was heartening to see this restraint and the avoidance of violence. Ironically, the tense but still peaceful standoff was day-by-day escalating the crisis exponentially – the restraints of the Army units refusing to confront the protesters head-on made the government seem weak and losing control. I was watching much of this development with a group of Hong Kong student friends, in our safe and comfortable campus – our emotions and fear for what was unfolding was palpable. We were students of similar age to the protestors and we also naturally empathised strongly with the protesting students' idealism. However, I was becoming increasingly angry with the student leaders who refused to call off the protests. I was getting increasingly anxious that the protests and occupation of Tiananmen Square must be called off now. After years of progressive reform, China and the future of Hong Kong was at the precipice.

The last week of May was dominated by the news in China – a crisis that could alter for good or bad the lives of over a billion people and the rest of the world. Compared to the struggles of people in many parts of the world, my life in England was a privileged existence, except I was also undergoing a crisis on two fronts. The first of these was not unique, there were three final exams to complete in that week – the stress of these exams was tremendous. The second crisis was unique and of partly my own making – I had allowed a deep muscle bleed to fester in my right forearm. If I had visited the local hospital earlier for a shot of Factor 8 treatment, there would have been no story. Amazingly, in almost three years of undergraduate study I had not yet visited my local hospital for treatment. Foolishly, I didn't want to break that record, plus I was still crazily stubborn about getting my bleeds treated quickly. As a result of this intransigence, my right forearm had become incapacitated with a deep and serious haematoma growing within the muscle tissues – the pain was searing and my right hand could hardly pick up a pen. The adrenaline of the finals and the desperate crisis in China somehow kept me going.

At the end of that week, in the late evening of the 3rd of June, the infamous Tiananmen crackdown started. Crack troops from distant provinces, with no ties to the local populace were sent in to clear the square and end the protests. Across the whole of China, all dissent was extinguished. Leaders of the protest were rounded up and detained. Hundreds of protestors, as well as soldiers, were killed. With a group of HK students, we watched in horror and dismay as tanks and soldiers took control in Beijing. All of us worried deeply about the future of China and HK, some were shedding tears, and one of the students exclaimed his wish for a revolution and the destruction of the Communist party. Holding down my emotion, I did not say much, but I definitely did not want to see a revolution in China or the demise of the ruling party because the alternative would have been chaos and even more human suffering. I followed the news intensely even though there were more final exams to tackle the following week.

In my own small way, I was also suffering and extremely anxious for my own wellbeing; my forearm had not settled down – the pain was now excruciating, with a tinge of numbness settling in[74]. At the time, I was almost incapable of even scribbling any words or formulae, but I was fortunate that the next exam was a multiple choice paper! With China and my own health predominating in my mind, I later learned that I had only just managed to pass that paper. Although, non-religious, my scientific rationality has never managed to rid the doubt I have about the existence of God. When one is well, it is easy to dismiss such thoughts, but as I have discussed before, when there is a particularly debilitating bleed as occurred to me quite regularly then, it was difficult to resist praying for comfort and (maybe miracle) support. I didn't really know who or what precisely I was praying to, but in that desperately worrying times I wish to believe.

By the end of the week, my arm was finally on the mend, which was a huge relief[75]. My final exams were over too – I was really glad but worried that my performance had felt like a disaster. The situation in China had also settled down.

The CPC government had reasserted authority and regained order. But at what cost to China and the aspirations of the Chinese people? My reaction to the 4th of June crackdown was sadness and shame. Sadness for the death and casualties of innocent people, and shame that China had to resort to guns and tanks to suppress civilian protests. Shame that arrested people were showcased cowing and being indignantly subjugated – how could we overseas Chinese expect foreigners to show us respect when China publicly humiliates its protesting citizens? Sadness and shame that the genuine progress initiated and engineered by Deng Xiaoping may have been all for nothing as China appeared to be regressing back to hardline policies and doctrines. But I also felt strong blame and shame on some of the protest leaders – as even before the terrible crackdown of June 4th, by the middle of May, I was very angry at them for not heeding the advice of the moderate and reforming party secretary Zhou Ziyang[76], their University professors and many ordinary people to cease their protests. If they had heeded them, with the return of normality and stability, the moderates' agenda of more progressive reforms may well have prevailed. Instead, I believe fame, ego and power politics corrupted some of the student leaders to take irrational and self-aggrandising decisions. One of the most extreme and arrogant leaders was Chai Ling, who a week before June 4 conducted an interview[77] with a US journalist expressing the view that bloodshed in the square would be desirable. I was livid when I saw that, as in my view, gradual and measured progressive reform is always preferable to wishful revolution, which tends to end in chaos and greater sufferings.

In the aftermath of the June 4th crackdown, China, which was previously the leading reforming Communist country, had turned into an international pariah. I worried that China might go through another phase of turmoil. Why was it that China seemed so fundamentally incapable of sustaining modern development? What direction would China go now?

## The Cambridge experience (and about Myjaugoek)

In the late Eighties, two of the most popular research subjects to study were cosmology and particle physics. I had fancied myself doing either of these subjects even though the competition for places was the most intense of all theoretical physics research. Indeed, as I attended interviews[78] for these courses, I found myself competing with other students who were much smarter or more knowledgeable than me. I wasn't offered these places which was just as well as I was getting rather sceptical with particle physics – it had become more like a classification subject due to the proliferation of elementary particles emerging from the great particle accelerators in CERN and elsewhere. As for cosmology, my grasp of Einstein's General Relativity was rather limited. I would have in fact loved to continue research in the foundations of quantum physics i.e. related to Bell's theorem discussed earlier, but at that time there was little funding for PhD study in the then perceived quasi-metaphysical field[79]. The three years of undergraduate physics was great, but it was clear that even three years of intense study was not enough to really appreciate some of the more advanced and fundamental areas of physics. Hence I wanted to learn more graduate level physics before considering PhD research. Cambridge University had one such course – the so called Part III Maths Tripos: a master-level taught course at the Department of Applied Mathematics and Theoretical Physics (or DAMTP as it was commonly known) which was then based near the city centre at Silver Street, near the famous 'Mathematical Bridge' of Queens College. The department was particularly famous as its head was Stephen Hawking, who was then basking in the glory of his surprisingly popular hit book 'A Brief History of Time'. On the course syllabus was the most exciting menu of topics – quantum field theory, general relativity, applied gauge theory, black holes, advanced quantum field theory, string theory, supersymmetry, quantum many bodies theory and so on. Wow, amazing subjects to be taught by world experts at the most famous University in the world. I was seduced, and I could get there if I got a First.

The wait for my finals results was one of the most anticipatory times of my life. I knew pretty much that I fouled up my finals, but my high scores in my projects and previous exams may just compensate for the deficit. I dared not go see the results for myself – instead a fellow physics student friend Tim came back to tell me "Kin, you will be punting in the river that sounds like your name". What? The river Kam! I jumped up with joy. Indeed, when I checked the scores, I had just scraped the First, narrowly avoiding a viva thanks to a single bonus point rewarded to me as a result of achieving top grades in two small assignments during my first year. This was a great lesson that every mark counts.

There was also a late offer from Manchester University to do a PhD in the phenomenology of particle physics – it was tempting, but the opportunity to experience Cambridge was just too great[80]. For anyone who has not attended Cambridge, there is mystique and awe about the place. Stephen Fry, the famous television personality put it most succinctly when he said "The best thing about having gone to Cambridge University was never having to deal with not going there".

Before I went there, there was one life changing skill which I had finally decided to learn. I was going to learn how to do self intravenous injection. Once I mastered this unnatural and obviously uncomfortable act, I could treat myself with Factor 8 at home at any time instead of the hassle of travelling to the hospital for treatment. My previous reluctance to acquire this skill had left me with numerous painful and at times dangerously serious internal bleeds in the muscles and joints of my body – although, I also felt that decision may have also contributed to my fortunate escape from HIV infections. The need to accept home treatment was really forced upon me by the increasingly frequent bleeds I was experiencing in my right knee. It had turned into a targeted joint[81] – previously I was able to tolerate the relatively mild pain and discomfort due to the resilience of youth, but lately it had become more debilitating and frustrating. The doctor had recommended prophylactic treatment which meant giving myself Factor 8 treatment three times a week for a period of several months – irrespective of whether or not I had a bleed. The idea was to keep a constantly sufficiently high level of factor in the body at all times so as to prevent any bleeds occurring, and in my case to help break the vicious circle (see diagram below) that inflicted on my knee and on my life.

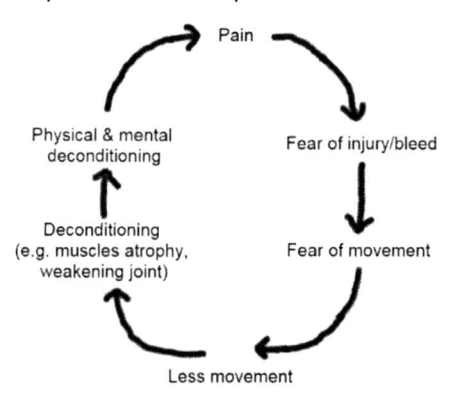

Pain → Fear of injury/bleed → Fear of movement → Less movement → Deconditioning (e.g. muscles atrophy, weakening joint) → Physical & mental deconditioning → Pain

Although an adult, I continued to see Sister Shaw at the Royal Manchester Children's Hospital who taught me how to self-administer the lifesaving fluid. Preparing the medicine was straightforward – just add the sterile water to the F8 powder in the bottle, stir the bottle until it becomes a clear liquid, and withdraw it into a syringe. But getting the F8 solution into my body was the tricky part as it needed to be injected into a vein. I learned the injection techniques quickly – I had after all watched and experienced countless injections by doctors and nurses, even though sometimes they did it appallingly. However, self-injection was really difficult psychologically; holding the needle and then consciously putting the needle into the vein, knowing that it's going to hurt. Worse still, my overactive imagination and anxiety made the process harder still, as I feared the needle missing the vein or over-puncturing it. I had no choice though – it was necessary to inflict pain to save myself from considerably greater pain and misery.

"Cambridge" – just speaking out that name conjures up a strong resonance of tradition, intellectual brilliance, elitism and privilege. I had never visited Cambridge before, so like many people I had images of toffs in gowns, dining at high tables in ancient colleges and of course posh students punting on the narrow river through picturesque colleges. Indeed, I recall being quite nervous on my first phone call to the college as I arranged to find accommodation in the new academic year. I felt the need to speak extra clearly and properly as though that was a qualification to be there! Fortunately, I need not have feared as on my first trip there, the people I met turned out to be quite 'down to Earth' and 'normal'. I was looking for a room in a shared house as all college accommodation was taken. The first house I visited was inhabited by an old and eccentric looking lady – she was kind and friendly, but the interior of the house was like her, a bit old and eccentric too. I told her I would be studying physics and she then just casually told me that her son was Brian Josephson who just happens to be a physics Nobel Prize winner! With great enthusiasm, she shared with me several enchanting stories[82] and pictures of her genius son. It would have been fascinating to live there, but I would have rather shared a place with other young students. The next house I visited was more appropriate – sharing the house with a group of other graduate students in the Cherry Hinton area of Cambridge. Later, as I moved in, I learned that the landlord, whose son was living in the house, happened to be a Professor in Haematology at UCL. What coincidence I thought – this network might come in handy although they were not aware of my condition in the beginning. That's the thing with Cambridge which most impresses me, there are just so many illustrious people around the place, and I felt rather privileged to have somehow got there. To study physics at Cambridge was like a dream come true. Many of my favourite legendary figures of theoretical physics had been there; Isaac Newton (arguably the greatest scientific genius of all time), James Clerk Maxwell (unifier of electricity and magnetism, which gave rise to the theory of electromagnetism), Paul Dirac (one of the founding fathers of quantum mechanics, quantum field theory and theory of antiparticles) and numerous others. Of course, world renowned physicist Stephen Hawking[83] was a common sight in the department. So there was great excitement to be learning there – amidst such great heritage. But the Part III course was a shock and big disappointment to my system. The style of teaching was completely different from my previous experience. There was virtually no physical explanation of the subject matter. As soon as the lecture began, the lecturer appeared to relentlessly scribble down an unending series of algebraic expressions and equations. All I had time for was to copy from the blackboard without any real appreciation of its physical or theoretical origin. Important equations or theorems just seemed to be plucked out of thin air, and if one was quick enough to follow the relentless algebraic derivation, then maybe just then, did I derive the pleasure of understanding a new subject. Sadly, every subject field was taught like this. The intensity and speed of delivery of the lectures were furious and ridiculously fast, but it was the lack of physical explanation in the lecture which was the biggest disappointment! Perhaps I was now the stupidest student among the crowd of top achievers, unable to comprehend fields which others took for granted. Indeed, some of the students, particularly the former maths Cambridge undergraduates seemed to handle the course just fine. Fortunately, for my own confidence, many other students from outside institutions felt the same as me. I reflected soon after that this cultural shock of learning was due to the following three reasons: there was a distinct difference in the teaching styles between physics departments which I had experienced, and those of the 'applied' maths departments such as at DAMTP. The latter tended to neglect physical explanations, instead favouring abstract mathematical formulation. Secondly, the advanced theoretical subjects which I was studying such as quantum field theory were fundamentally inspired by mathematical consistency, beauty and conjecture, instead of concrete physical principles. At best the origin of quantum theory may be explained by abstract analogy only. The historical reality is much of the derivation of quantum physics is ad hoc, and yet somehow fits and successfully predicts experiment observations brilliantly. Finally, the expectation to simply understand the subject just from attending lectures could no longer be expected – for such an advanced topics one must expect to do a lot of self-study. When I realised this, I felt less disappointed by the teaching – instead I became rather disillusioned with the subject itself which no longer seemed to have an easily comprehensible physical model or basis.

I was also disappointed at not getting a college accommodation. Living in a shared house at Cherry Hinton wasn't ideal because it was quite far from both the department and the college. Unlike the cosy living at the York University campus, I had to commute by car as my fitness to walk long distances had been severely hampered by my bad right knee. It was quite stressful to find sneaky free parking spaces near the department every day. I was too proud to seek special parking privileges for my condition. At that time, I still thought of myself as 'normal' and was even more reticent to divulge my H condition because of my own sensitivity about it, and the stigma attached to it. So overall there was a bit of adjustment initially which included attending Saturday morning lectures, and tutorials that finished after 7pm! Cambridge at night is eerily quiet, although there is something especially atmospheric about that. I would not recommend Cambridge for the night life, but it's just fine for academic inspiration.

The Part III maths course had a reputation of being notoriously hard, with folklores of students falling ill under the stress of studying it. Unfortunately, I had even more stress than most – I found the act of self-administered intravenous injection psychologically very difficult. As useful as home treatment was, I almost could not accept this new routine in my life. But I had to try to get my knee in a more stable condition as I was tired of living with increasing bouts of acute debilitation. But despite thrice weekly prophylactic treatment, the joint remained problematic and almost permanently inflamed. Physiologically, the treatment helped somewhat, as without it, the knee would no doubt have incurred more severe bleeds and subsequent debilitation. That wasn't good enough though – the prophylactic treatment should have stopped and prevented all bleeds in my joint, so it could recover properly. But to try to live normally meant I didn't give the knee the proper rest it deserved whilst I needing to commute and attend lectures 6 days a week. Somehow, I managed to soldier through November – an eventful month I recall, with the peaceful fall of the Berlin wall[84].

I don't like winter because the cold often aggravates my condition. Cambridge can feel especially chilly as the bitterly cold Eastern wind blows through the flat plain of East Anglia unhindered. On one such day while visiting a Part III student acquaintance at Trinity college (where the great Isaac Newton once resided), my knee swelled up severely. There had been early warning signs, but I ignored it until it was too late – I could hardly walk at all. I limped my way through busy Cambridge city centre to my car at DAMTP in Silver Street. This was one of the handful of walks in my life that felt like I was in an epic struggle. When I got inside the car, I collapsed on the seat; relieved, but in desperate pain and worried for the state of my knee. After a short rest, precariously, and with much concentration, I managed to drive back home using the action of my ankle as my knee was mortified by the intense swelling and pain. Using fixtures and furniture in the house, I managed to retreat to my bedroom where I immediately prepared and gave myself a dose of Factor 8. The situation was serious, as the whole right knee had turned into an oversized cricket ball. It was red hot, turgid, vulnerable and absolutely sensitive to the slightest movement. As usual, when it came to my condition, I kept myself to myself. There was just me to rely on. The next day, there was no improvement, plus never seen before red rashes had emerged all over my body. An ambulance later ferried me to another of Cambridge's famous institutions, Addenbrooke's hospital. Thoughts ran through my mind that skin rashes were often associated with symptoms of AIDS.

I was hospitalised. A haematology consultant briefly assessed me, followed by an orthopaedic consultant. X-rays were taken, and a blood sample taken for an HIV test. It had been several years since I was hospitalised, a place which I have always preferred to avoid, but also at the same time, felt very relieved to be there. There was nothing new about my treatment though– the same Factor 8 was given, plus total bed rest which I could not do before. The latter made a big difference as I didn't have to struggle to do the simplest of things. I was surprised to see how quickly my severely swollen knee reduced in size day by day. There was both good and optimistic news too – the HIV test had returned negative. I don't recall if the red rashes were ever diagnosed, but they cleared up anyway – they were probably caused by a reaction to a bad batch of Factor 8. And also the X-rays showed my right knee joint apparently didn't look too bad, despite years of bleeds and inflammation within it[85]. But there was also a big realisation that I hadn't really appreciated until then. The boring but peaceful bed rest allowed me to reflect that I had not been able to run or walk briskly for a long time. For the first time in my life, it suddenly dawned on me that I had a permanent disability. Why had it taken so long to realise this obvious state of being? Being an H meant I had regularly experienced infirmity and disability from almost all parts of my body, but they were usually short-term. Even for my right knee, which had been given me trouble ever since my devastating psos bleed over a decade earlier, I had attributed the problem to my inability to break out of the vicious circle of recurring bleeds. Previously, I thought if I could break out of it then I would be normal again. The reality was that recurring bleeds had over time insidiously caused significant cumulative damage to my joints. My naive and undeserved feeling of youthful invincibility was finally shattered[86].

I really wanted to break out of the vicious circle. Perhaps the doctors at this renowned hospital could explain and resolve my knee problem. But none of the consultant doctors who had seen me briefly on my first day came round again until a week later. I was really hoping to have some time to discuss with them my problems, but instead they appeared hurried and after only a brief look said I could be discharged that day. I was grateful that my knee had managed to return back to a healthy shape so (miraculously) quickly. In fact, I hadn't seen it looking that normal for such a long time that I had accepted the puffy and slightly inflamed state as my standard of normal. The hospital stay demonstrated to me that with sufficient control and appropriate rest, it was possible to have the leg return to almost normal shape, if not function. But I wished I could have been allowed to stay just for a few more days, for it to settle down more firmly. I was very disappointed that I didn't have the opportunity to discuss at length my situation with the doctors. In the week as an in-patient, they had only visited me once, and even then the encounter was fleeting and rushed. To my mind they discharged me prematurely, without properly understanding my condition. The surprisingly healthy look of my knee didn't last long though. Within a couple of days of my return to the shared house, the knee flared up again. I felt angry as this would not have occurred if the doctor had bothered to talk to me before giving the discharge order. I became even more frustrated as there seemed to be no respite for my condition. I treated myself aggressively over 3 or 4 days to try to regain control. The treatment itself was becoming more difficult too as the veins on my arms were bruised and battered from the recent series of repeated injections.

Somehow, I managed to just get 'well' again. The Christmas holiday break had started and I drove back to my parents' home/shop for a well-deserved rest. But it wasn't long before the knee flared up again and I ended up lying on my bed for most of the Christmas holiday. After years of living with the inconvenience of H, the last few months of incessant and intensified setbacks were too much. In a depressed moment of frustration, anger and hopelessness, I couldn't hold my emotions anymore – in the presence of my mum I wept and cursed profusely at the pain and misery of my frustrated life, the fate of living with this fucking disease. I exclaimed repeatedly "WHEN will my bloody knee give me a break". I was tired of living like this, in a permanent and oscillating state of severe debilitation followed by quasi-wellness and then back to more disability, and so on and on… At desperate moments, I thought fleetingly about chopping off Myjaugoek ('jaugoek' meaning right leg in Cantonese).

Although dark thoughts passed through my mind at the times, the world is too interesting to drag me to the abyss of permanent despair. In bed, I watched avidly the news of another revolution unfolding; the fall of the Ceausescu regime in Romania, and rumours of the imminent release of Nelson Mandela. China was still in lockdown mode. The suffering of others somewhat nullified my own misfortune[87]. I returned to Cambridge in the Lent term, refreshed and re-stabilised. While Cambridge in winter is gloomy and somewhat austere – under the warmer air and lighter skies of spring, Cambridge purveys youth and optimism. My mood had changed too; Myjaugoek was under control and I could enjoy being a more ordinary student again.

I made it through the summer term and its torturous exams. The Part III maths course had not helped me to decide on my future area of study. On the contrary, two subject areas which I had previously contemplated would not be suitable anymore; Cosmology and Einstein's General Relativity were both too grandiose and difficult, while the topic of theoretical particle physics (which uses quantum field theory) was too weird and abstract. In fact, I felt tired of studying. It was time to find a job.

# 6. WHAT IS LIFE?

## The first day

The transition from full-time study to full-time work is a major milestone of one's life. My plan was to get an interesting graduate level job and gain some experience by working there for a year or two, before I returned to academia to study for a PhD in physics, in a subject which I was passionate about and not beyond my ability. Ultimately, I hoped to settle down in an academic job doing teaching and research.

In the early 90's, the number of good graduate position jobs advertised was relatively numerous. The chance of getting the job was probably easier too because the pool of graduates was much smaller then. Hence, within a matter of a couple of months, I had three job interview offers; at IBM (as a software developer), at the patent office (as a trainee patent examiner) in Cardiff and at Culham laboratory[88] as a 'Theoretical Physicist' working in the field of nuclear fusion research. Although I had not particularly wanted a physics related jobs, the last position was by far the most interesting and noble because the ultimate goal is to produce unlimited clean energy via controlled nuclear fusion processes. The fact it's directly related to applied physics was a real plus as I had grown weary with the crazily supercharged mathematical physics at Cambridge. Coincidentally, the first Gulf war had recently started, and I recalled citing that in my interview as a further reason for nuclear fusion research.

To my great delight and amazement, I was offered the job after two friendly but testing interviews. I felt amazed because I had relatively little experience of research compared to most of the other candidates who held at least a PhD. Furthermore, it was a great relief that I got the job because during the interview process I had to reveal my medical condition to the personnel department which I worried may have jeopardised my chances[89]. Hardly any friends and acquaintances knew of my H condition, so having to divulge it to strangers was personally very uncomfortable to me.

Unlike some people who waste no time between jobs, I gave myself a one month break before starting my first working day in the beginning of December – after all, there was not going to be much opportunity to truly feel free from responsibility once one starts a job. Also, I was by then quite obsessed with the state of my right knee. I wanted to build up its wellbeing by subscribing to a proper prophylactic treatment regime and stringently adhering to the taught physio exercises to build up the quad muscles. The point being that the greater the muscle strength in the quad the less vulnerable the knee was to injury and bleeds. Unfortunately, all I could manage was simply to keep the situation from escalating out of control – by this I mean not suffering bleeds sufficiently bad to debilitate me – most of the time I just got on with life bearing a tolerable puffy knee and a resigned 'normality' of living with a lame leg.

Culham laboratory is situated in the Oxfordshire countryside in Abingdon, which is about 5 miles south of Oxford. It is also the site of JET (the Joint European Torus) which is the world's largest fusion experiment machine. Around this region are also several other world famous physics laboratories; Harwell laboratory, Rutherford Appleton laboratory and more recently the Diamond Synchrotron laboratory. I felt really privileged that I would soon be working amongst them. I recall years earlier, as a schoolboy, I read about the famous JET experiment and literally dreamt of working as a nuclear[90] physicist. Soon, my literal dream would come true, even though I had not planned it!

Such was the number of employees working within these research laboratories, the employer UKAEA had a dedicated hostel for new staff and visitors, called Rush Common House near the centre of Abingdon. Full of expectation and excitement of my first proper job – the sense of exploration growing as my white Ford Escort made the three hour journey from my parents' home in Manchester to my new beginning in the south. It was a memorable drive listening to my favourite Bob Marley cassettes and at times catching on the radio the hauntingly beautiful Chris Isaak's song 'Wicked Game'. Rush Common House was like a college dorm – living in your own basic bedsit with its own wash basin while everything else was shared amenities. There was also a dining hall that served breakfast and dinner just like college, except the people are more mature and quieter than boisterous undergraduates. I rather liked this as it afforded me more privacy than living in a shared house, and it reminded me of the collegiate feeling I experienced at York (ironically, I missed much of that famous Cambridge college experience as I didn't get a college room). I hoped that I might make some new friends too.

Like most people feel on their first day at work, I was nervous and anxious about being introduced to new colleagues, asking a lot of questions about work, and making small talk whilst trying to sound intelligent and wise. At one point during the day I ran out of meaningful questions and babbled out an irrelevant and random thought to the department's boss Jack – what's the overtime rate? Whoops! That was really embarrassing as scientists are supposed to be uninterested in money. We do science for the love of it! I was shown my own office which I would be sharing with another new colleague starting in the New Year. There was a blackboard for each desk, the ubiquitous tool for the theoretical physicist – there was a sense of pride that having this job had confirmed me as a verifiable physicist. There was also a bit of walking around that day, being introduced to various places like the library and the experimental laboratories. But something else was starting to dominate my mind. My right knee had started to swell up and it was getting increasingly uncomfortable and stiff. As I was guided to the library, I tried my best to walk normally as I hated to be seen differently, but it was obvious that I was by then walking with a significant limp. Fortunately, the library afforded me an excuse to sit down as I pretended to be concentrating on the books and literatures in the library. I CANNOT believe this; on my very first day at work, I was debilitated by a bleed. The anxiety, worry and embarrassment was truly overwhelming as I feared arousing concern and bewilderment from my new employer, as well as possibly becoming too debilitated to go to work the next day. I kept my knee really still during my allocated time in the library, hoping it would not get worse. Eventually, with some difficulty, I limped back to my office. I worried what my new colleagues thought of me as I went from normal in the morning to now limping obviously. I was glad to be sitting down again, putting on a normal face; I went through the motion of signing various administrative forms and reading through the handbooks while all the time worrying gravely at the state of my being. I was desperately hoping for the end of the day.

After what seemed like an eternity and a heroic struggle to get to the car, I rushed back to Rush Common House and immediately prepared and infused myself with a larger than normal dose of Factor 8. Also, I wrapped my leg in several layers of bandages for stability and compression while trying to be careful that it's not put on too tightly or the situation could worsen. I made myself some pop noodles that evening as I dared not inflict further stress on the knee in getting to the dining hall. I laid on my bed for the rest of the evening praying the medicine would work its magic so I could go to work in the morning normally. I didn't have much sleep that evening.

When morning came, the swelling in my knee had reduced significantly. I gave myself another dose of treatment before going to work. The rest of the week went without incident, and I vowed to prevent this type of mishap from happening again.

## Nuclear Fusion

Most of us are aware that a significant proportion of the power generated in the UK is from nuclear power stations. However, not everyone will know that all the nuclear power in the world is generated via the process of nuclear *fission,* as opposed to nuclear *fusion.* The former involves splitting heavy atoms into lighter elements, whereas the latter involves combining (i.e. fusion) two light elements to form a heavier element. It can be said that life on Earth is due to nuclear fusion, since nuclear fusion processes are what makes the Sun shine. In both processes, the extraordinary amount of energy comes from the conversion of the missing mass (i.e. the difference in the 'before' and 'after' masses of the reactants and the final products) into energy, which according to Einstein's famous $E=mc^2$ equation shows that even very small amounts of missing mass (m) can lead to huge amounts of Energy (E) as the speed of light (c) is such a very large number[91].

The current crops of nuclear fission powered stations are controversial because fission once started is fundamentally unstable leading to 'runaway' reactions which have to be controlled very carefully. Its by-products are also extremely radioactive with long half-lives. Although very rare, the Chernobyl disaster of 1986 and the more recent Fukushima nuclear disaster in 2011 are two unfortunate examples of the dangers of nuclear fission power generation.

On the other hand, a controlled nuclear fusion powered station would be fundamentally safe because initiating and sustaining the nuclear fusion process requires very precise control of many parameters. Any disruption from the optimal running parameters would simply lead to the reaction to shut down. Hence, even today there are no nuclear fusion power stations because the controlled nuclear fusion process is so VERY hard to start and to sustain. The reaction which is most likely to yield fusion is between Deuterium and Tritium[92] which yields extremely high energy Helium and neutrons, whose energies are harnessed for energy production, as well as contributing to heating the fuel itself. Note, despite their exotic names, Deuterium and Tritium are just heavy forms of hydrogen atom which is abundantly available from sea water.

At very high temperatures, the atoms of Deuterium and Tritium are completely stripped of their electrons forming a soup of positively charged ions and negatively charged electrons, known as a plasma. To achieve an efficient rate of fusion requires the plasma to be heated to 200 million degrees Celsius (about 10 times hotter than the temperature at the centre of the Sun!). This requires the plasma to be confined and heated from an external source for sufficiently long enough time such that the released fusion energy is significantly greater than the consumed energy to sustain the reaction. The astute reader may have spotted the challenge of how to contain something so hot without the container getting damaged? The answer is that these very hot charged particles are confined by magnetic fields which force them to move in a spiral path. Furthermore, to minimise losses of hot plasmas, the favourite design for confining the plasma is a system that is shaped like a ring doughnut (a torus) with appropriately configured magnetic fields to restrain the plasma fuel from touching the walls of the reactor[93]. Hence much of fusion research around the world and in Culham is centred on experimental devices with a torus structure (also known as a tokamak), e.g. the Joint European Torus (JET) was and is still currently the largest and most successful machine to study nuclear fusion. The scope for research is very broad, covering plasma physics, nuclear physics, atomic physics, material physics, radio-frequency wave heating physics and numerous related engineering issues.

One of the major challenges of creating a successful fusion reactor is to tame the instabilities of the plasma – as I recall by a Culham scientist (I think it was Tom Todd) explained it thus: "...keeping the plasma under control via magnetic fields is like trying to hold steady a wallop of jelly with a bunch of elastic bands". At the theory division in which I was based, I was very privileged to work on some of these problems in probably the best theory group of plasma and controlled nuclear fusion physics in the world. There were some great scientists there such as Jack Connors, Jim Hastie and the late Chris Lashmore-Davis who was my first boss. Also there at the time, were Chris Bishop who later became Chief Research Scientist at Microsoft Research UK and Bryan Taylor[94] who had a number of plasma related theorems named after him.

There were two other new young recruits in the theory department at the time of my appointment. They were both Scots, Colin Roach who had a PhD in experimental particle physics from Cambridge, and Brian Harvey, previously a postdoc in plasma physics at St Andrew University. Brian would arrive in the New Year while Colin had already started at Culham a couple of weeks before. I felt immediately at ease with him with his open manner and ease of laughing out loud. Brian and I were to share an office together, so I hoped he would be a nice guy. Fortunately, he too was a very friendly person, except for his intensely reserved and reticent streak, which was nevertheless easily broken provided I initiated the conversation. While Colin spoke with a distinctive but easily comprehensible Scottish accent (which had no doubt been toned down by his years of living in Southern England), Brian's Lanarkshire accent remained original to his roots – hence with some bewilderment it took me the best part of three months before I could easily understand him! Besides physics, we shared similar interests too, like cars, technology and science fiction, especially Star Trek. Amusingly, he told me that he did research in plasma physics because of Star Trek – the starship Enterprise being propelled by a plasma drive. I would not be surprised that if anyone were to make a breakthrough in this area, he would be the one. When it came to solving physics problems, the expression "sharp as a hot knife through butter" really comes to mind. I have never met anyone whom I was close to that possessed a physics mind as sharp as his. We talked a lot about cars too – for some reason, when I first saw the red Ford Escort XR3i parked at Rush Common House, I knew that it belonged to him even before we were introduced. As the most junior member of the department with the least experience in research and plasma physics, I felt very lucky to share the office with Brian as he was especially open to helping me out or to discuss my physics problems.

On Saturday 9th November 1991, fusion research was in the global news – for the first time in the development of fusion energy, JET managed to produce 2 Megawatts of power in a pulse lasting 2 seconds. It was the first time that the optimal 50:50 mix of Deuterium and Tritium fuel mixture was used. Even though I wasn't involved directly with the JET project (which was just a few minutes' walk from my office), I felt a real sense of history and purpose to be part of this great community.

## The relentless vicious circle

Although I had settled into my job well, the problems with my right knee were persistent and unrelenting. In my first couple of years at work, I kept my condition just about under control so that I was performing my work duties and not taking any more sick leave than the average person. But, I felt increasingly frustrated and depressed of living on the edge – every other week it was like living under the situation I described in my first day at work. On the surface, I was working and surviving but internally I was constantly in a state of uncertainty, discomfort and fear of debilitation. Inevitably my mind was restless. Instead of thinking physics, I was thinking of the knee. Instead of going out on the weekends, I was held back by the knee. Instead of getting on with life, the knee is my life.

Just occasionally, I would go through a week or two of just feeling fine – of just being normal. I felt great. But the cruelty of H is that a bleed can quickly debilitate with no warning or mercy. This frequent and periodic cycle of 'normal' followed by profound infirmity makes H especially wicked because it gives you a taste of normality before cruelly snatching it away. In my darker days during this period, I thought once again about amputating 'that' bloody leg. Anything would be better than living like this.

During this time in the early 90's I had two major health status changes. The hepatitis C virus (HCV) test had become available and I was tested at the Churchill hospital in Oxford around late 1991. The test results confirmed that I was HCV positive. I wasn't surprised and remained ignorantly sanguine about it. After all, there was very little information about how serious it was and I had been living with it asymptomatically for many years already. I didn't let it bother me too much except I would have to be extra responsible with any spilled blood or if I met a partner. Following this test, the hepatologist nurse who was also in charge of disability benefit entitlement insisted that I apply for disability living allowance which would entitle me to a disabled parking badge. I was very reluctant to apply for it as I felt too proud to become a benefit receiver as well as having to publicly admit that I was disabled. But, I relented because the disabled parking badge would be very handy now, given my increasingly limited and unstable physical condition. At the age of 25, I finally conceded that I was disabled; not just a normal abled-body guy who was sometimes infirmed!

Year three at work came quickly... not because that time passed quickly because of having a great life, but rather that I had settled down to a routine of life involving going to work, being at work, back home, eat then rest on my bed. The latter I did a lot because of the knee. In fact, come to think of it, I spent a substantial amount of my time lying down and resting my knee during much of the previous 10 years. I conjecture this may be why I am the tallest person in my family as gravity has had less time to constrain my growth! 1992 was a truly reflective and difficult year as the culmination of damage to the knee was making my life hell. I recall being bed bound in my ground floor bedsit in East Hagbourne, Didcot for almost three weeks. I remember this period well as it coincided with the debacle of Sterling leaving the ERM[95]. I called work to inform of my sick leave, but I didn't tell anyone else of my situation as I did not want to bother anyone. No one came around as none of my family, colleagues or few friends knew about my real condition. The other tenants tended to mind their own business, so the only person that I got to talk to were the delivery men from the local takeaways, collecting the food through my ground floor bedroom window as I could not get to the door. For some stupid reason, I did not own any crutches at the time[96], so even getting to the nearby bathroom was a struggle, as was getting to the kitchen which I could only reach with the use of a chair as an improvised Zimmer frame. If I had to move in my bedroom, I crawled on the floor to avoid bearing weight on the severely swollen knee. Also, at one time even giving myself the treatment became a battle of will as the two favourite injection sites on my arms became bruised and overused. I had concern that one day I would not be able to treat myself anymore as all my veins became too damaged. Even though I withstood all these adversities well, it was a lonesome and stressful existence. In the past, my main concern during these difficult periods was the fear of missing classes. But now my main concern was the responsibility to others at work and about my future – health wise and career wise.

## Natural Mystic

Like so many times before, I went through the usual anxious and frustrating time of waiting and eventually going through the gradual recovery process. There was much time for contemplation. When will I ever get better? What can I do to break out of this vicious cycle? What is the nature of the plasma instabilities in the model?[97] Why do I have to suffer so much? What is life?

The absence of the internet or mobile phones then meant these were lonely and isolated times. My only companions were the television, radio and a small collection of books and cassette tapes. Amidst these lonesome times were the songs and music of Bob Marley who gave me much inspiration and hope for better times:

> *One good...*
> *When it......*[98]
>   :
> *From the track* ***"Trench Town Rock"***

There is always hope. The world is too interesting a place to lose hope and the joy of living.

Ever since I bought the timeless 'Legend' album at 18 years old, the music of Bob Marley has always resonated and uplifted me. In fact, my first memory of Bob is seeing in a news report a diminutive Black man with wild hair, singing and dancing in a trance-like state whilst uniting, joining and raising the hands of two caucasian looking men in a concert[99]. It left a powerful image in my mind even though I had no idea who he was or what his music was about.

His songs about freedom, struggle against inequalities, love and liberation are universal even if his context is different from mine. Hence, I dedicate this section to some of his songs which have so inspired me – in good and bad times. His songs speak powerfully to our soul with our own individual circumstances and destiny. Below is a selection of some of my favourites:

> *Good Friends..*
> *Oh good......*
> *In.......*
>   :
> *Everything's...*
> *Everything's...*

*I say.....*
　:
*From "**No Woman, No Cry**"*

This famous track first got me into BM as a young teenager. I wasn't really in tune with the man-woman and love-loss aspects of the song then, but the rhythm and above words so summed up the feeling of loss and yet there is still so much hope. The 'loss' here refers to my loss of being normal – my decreasing mobility, the loss of the freedom to roam, and be carefree about travelling the world.

　:
*No chains.......*
*I know........*
*And.......*
*And........*
*Still.......*
　:
*From "**Concrete Jungle**"*

This broody song was in the brilliant 'Catch a Fire' album which propelled BM and the Wailers to global consciousness. I like this song, instantly identifying my own past of living in the Concrete Jungles of Kowloon/Hong Kong and then that of the Sholver council estates up on the bleak Moors of Oldham. On a different tone, the words above describe so literally my loss and predicament. But just feel that last sentence!

　:
*Emancipate....*
*None but......*
*Have no.....*
　:
*From "**Redemption Song**"*

This is a beautiful song with just the acoustic guitar as the accompanying instrument. The differentiation and assimilation of cultures of the past and present is not an instant or an easy process – it requires an opening up of one's mind to truly appreciate that all cultures have good and bad aspects. Don't be trapped by our past, but don't be fooled by the present either. Let our minds adopt the best of our cultures. I do not fully get the last line about 'atomic energy', except perhaps that is like some strange new ideas (or cultures); one should not fear it.

:
*There's a......*
*I won't...*
*If you......*
*There's a......*
*Such a......*
:

From "**Natural Mystic**"

For me, BM is a natural mystic and a prophet. He had little formal education, yet somehow came up with words and music so powerful, so mystic. In my period of bad time, this song inspired me to think of miracles and a sense of spirituality that might deliver me to another world, a kinder world. Many of BM's songs have this ethereal quality, often citing the power of Jah (God) and of delivering justice. Although I am non-religious and was very uncertain about the existence of God, the words and rhymes of the following two songs have frequently inspired and uplifted my soul:

*:*
*Feel..*
*Lord...*
*:*
*Feel..*
*Lord...*
*:*
*From "**Put It On**"*

As an agnostic, I don't think I have ever said aloud in my adult life 'Lord I thank you' or similar phrases involving the word God. However, in the songs of BM and some other songs by others, these words resonate powerfully with a feeling of gratefulness – that despite all the difficult times that one has endured, I 'feel alright now' and I 'feel the(m) spirit' of living; Oh, 'Lord I thank you'.

*You're........*
*You're........*
*You're....*
*But......*
*Can't....*
*Can't....*
*:*
*Every man......*
*Who feels....*
*:*
*From "**Running Away**"*

In good times, my body helps me to move, function and explore the beautiful world. But this very same body can also turn into a burden and source of misery as happened so frequently, hence this line *"But you can't run away from yourself"* felt so poignant to me. But listening to the last two lines somehow lifted my spirit.

Most of Bob Marley's songs were written within an environment of hardship, poverty, prejudice and oppression. There was a song which was written under a different kind of suffering – the kind that I was most familiar with – the battle with one's failing health.

:
*When the.....*
*It means......*
*When......*
*You find........*
*Then......*
*Are you.....*
:
*From* **"I know"**

There is in the lyrics of this song, an undertone of resignation to one's fate. But as is true in all of his songs, even in the midst of one's greatest battle, his voice and music pass on the message of hope and salvation. I adored this song even before I realised it was apparently Bob's last song, written while he was gravely ill.

BM also wrote many love songs. Even though romantic love was lacking in my life at the time, the longing of love and of better times is apparent and even penetratingly obvious from these songs. I adored: 'Hurting Inside', 'I'm still Waiting', 'Lonesome Feelings' and 'Waiting in Vain'.

Finally, BM also made some great political and social awakening songs like the two tracks below.

:
*Every man........*

*And in......*
*So........*
*Cause that's.........*
  :
*From "**Zimbabwe**"*

Although BM wrote this great song specifically to celebrate Zimbabwe's Independence, the song expresses a strong conviction of solidarity and of fighting against injustice everywhere.

  :
*How good......*
*Before....*
*To see......*
*As it's...*
*Let it...*
*We are......*
*We are......*
  :
*From "**Africa Unite**"*

This is a truly aspirational song, calling for the unification of all Africans. More broadly, I heard this as a calling for all humans to cooperate and treat each other fairly, unite irrespective of race, background, disability and creed.

## God and self realisation

"Life is hard, then you die!"

I would never say that, although I was hearing it quite often from my office partner and dear friend Brian. He said that in jest and in an ironic way. I forgot what prompted him to say it, but I thought it was quite funny, in a ludicrous kind of way. Certainly, I thought my life was hard, although I was too superstitious to dare suggest such fatalism! I liked Brian a lot. In fact, I really missed him when he went away on a research assignment in Paris for six months. Brian was quite a strange guy; he is helpful and kind, and always responds enthusiastically to my questions or conversations, laughing out loudly when as I do frequently, said or did something absurd or stupid, and as always glitteringly sharp at almost any maths or physics problems I threw at him. But if I didn't initiate the conversation he would never initiate one! At one time, I got so annoyed with him that I decided not to initiate a conversation first.

After one week of total silence in our office, I gave up!

Now, I realise he was probably on the Asperger's scale, although I didn't know that then. Still, I love the guy; there was an innocence and purity about him that I really respected. But after his return from Paris, there was a change in him. He seemed more solemn and became less jolly than before. I heard that just before and during his time in Paris, he had suffered several bouts of dizziness and muscle weakness. As he was an extremely reticent guy, he wouldn't discuss the matter with me. I felt very concerned for him.

My own situation wasn't great either as 1992 came to its final months, my right knee caused me further grief again. This time I became an in-patient at Churchill Hospital, Oxford. I shared the ward with an older H-patient who had significantly more severe joint problems than me, one which required a joint replacement. All the frequent bleeds into his joints had led to permanent joint damage akin to someone with severe osteoarthritis. Just as I had made the late realisation at Addenbrooke's Hospital, Cambridge that I was disabled, now I realised that my fate may become his in a few years' time – unable to work full time and permanently disabled with severe mobility even on a good day. Despite this rather unsavoury prognosis, the experience of sharing each other's travail was an especially liberating experience for me as it was probably my first time of sharing the H-experience with another H. Also, he had a loving wife which gave me hope that in the future, I might have such support too – in direct contrast to my time at the ward when no one came to visit me during my whole time there. Yes, I felt a bit sorry for myself, even though I was used to handling crises on my own.

My stay at the Oxford hospital, like in Cambridge felt very impersonal and rushed. The senior consultants were only seen briefly, and that happened only once a week. They always seemed annoyingly unavailable and incapable of giving the patients just an extra few minutes of their time. There was hardly any opportunity to probe them with concerns or exchange additional information that would help them and the patients. Perhaps I was spoilt by the excellent kindness and patience I experienced at the children's hospital in Manchester. After what was probably about 10 days in the hospital, I was well enough to return to my digs in Didcot.

Several new tenants had recently moved into the house; a couple living in the largest room with direct access to the back garden, and a stocky Irish builder in his thirties on the first floor. The builder guy wasn't particularly friendly and had a rather mean demeanour. He was also frequently visited by a woman, who appeared concerned and caring of him. I thought she may be a family member rather than a girlfriend. Anyway, I didn't care too much other than trying to be friendly whenever our paths crossed. But a few days after my return from the hospital, I was to experience one of the most traumatic events of my life. While still sleeping around midnight in my bed, I was woken by violent banging on my bedroom door. Fortunately, as I always do, I had locked the door before sleeping for peace of mind. It was the builder who was then shouting hysterically at me, while violently kicking and hitting at the door. I had woken to a nightmare! An hysterical man was accusing me of saying something behind his back, and was now attempting to kick down the door. I reasoned that the door would not last long under his persistent kicks, and that when the door did give in, he would rush in to attack me. I needed to calm him down as quickly as possible or the consequences would be dire. I spoke to him behind the door in a calm and soothing tone, assuring him of my innocence and trying to calm him down. To sooth him further from his rage and continuous barrage at the door, I made a quick but possibly very wrong decision to tell him that I would open the door.

Gently, I unlocked and opened the door.

He rushed in, face flaming red with a madman look and fuelled with alcohol smell. He grabbed me by the throat. His thumb and fingers gripped hard on it while my instinct for survival stepped in, which was to back off rather than to resist because I knew instinctively that any form of resistance or even defensive block in that situation would immediately lead to an even more violent reaction. I moved backwards while trying to be calm and trying to pacify him. His powerful hand now lost the grip on my throat, but he was clearly still moving towards me as I continued to show subservience to him, until I was forced backwards to my bed, where I immediately sat down on it to demonstrate my now even less threatening position. He came forward still, with a demented look, his face now just inches from my face, a barrage of raging words and spit flew in my direction. I continued to try to soothe him with calm words and body language. At one point, he even poked me in the eyes which were thankfully protected by my spectacles. Thoughts came close to my mind that perhaps I should fight back. In fact, I had prepared for it. Before I opened the door, I had put one of my steel Chinese acupressure balls[100] in my pyjama pocket. I thought briefly of taking it out and subduing him with it.

Very fortunately, that wasn't necessary. I knew if I did, there would be no going back. If I failed, he would most likely have killed me. Miraculously, I succeeded in calming him down and he left the room abruptly. The whole episode took perhaps no more than five minutes, but it had felt like an eternity. I didn't sleep that night. It was one of the scariest events of my life, threatened and assaulted by a maniac in my own bedroom. Beside the shock and upset, there was an aspect of this incident that lifted me; that I had the instinctive calm and aptitude to subdue this violent and threatening situation, peacefully. I managed to go to work the next morning, although as a form of self therapy, I could not resist telling my close colleagues about the incident. The incident was reported to the police, but no further action was taken because I bore no signs of injury. Miraculously, the madman's powerful grip on my throat hadn't caused any bruising and even the deep red marks he left initially in the assault had cleared away. That day, I sought new accommodation immediately as I didn't want to live with such a dangerous and violent individual. In the days following the attack I discovered that whilst I was in hospital he had made a similar threat to another tenant who had fled the house before my return. Had I been told of this incident, I would not have opened the door voluntarily and tried to negotiate with him. He could have been carrying a weapon which could have easily led to an entirely different scenario.

After I moved out, I found out that the woman who visited him frequently turned out to be a social worker. Apparently, the builder was a schizophrenic living under the 'care in the community' scheme. I was furious when I realised this. None of the so-called sane people or organisation had the decency to forewarn innocent people of their dangerous experiment. For a significant period this violent incident psychologically scarred my mind. I became very heighten to the potential of violence, as well as more easily angered by injustice. I felt more prone to aggressive thoughts in situations where I had been wronged – real or perceived. If there were anything positive that came out of this incident, it was a greater comprehension of why violence begets violence – of why previously sanguine adults or even children, who have been subjected to violence may themselves become perpetrators of violence.

I moved back to Abingdon, this time sharing a nice bungalow with just one other person, a mild mannered professional engineer who was also the landlord. This house felt peaceful. And it was in this house that I was to gain one of the most profound experiences of my life.

Life is a struggle. It's a relative term. Some complain over trivial things, and some live quietly and determinedly despite the hardest of hardships. Sometimes, the feelings are made worse because we compare it with others. As a H, I am aware that my life has been much tougher than most, but still luckier than many. With a rebellious attitude of quasi-denial of my H status, I worked hard to achieve a normal life, get a job and move on. In some ways I have achieved that. But in the years leading up to my 26th birthday, the relentless vicious circle concerning my right knee had brought me low too often. The frequent and repeated cycles of being severely infirmed periodically before returning to normality is probably unique to people with H. In my youth, I often thought that severe H's not receiving proper treatment must be like boxers who have to endure their injuries from fighting every other week. Hence, I have developed a much more heightened appreciation of what it feels like to be normal and able-bodied. Normal and able-bodied people often just take their good health and mobility for granted, until they are laid low by injury or disease.

How blessed it feels just to be healthy!

Except frequently now I felt exactly the opposite. The bad days are more intense than before and the normal days were becoming fewer. Beneath my stoic appearance was an increasingly anguished and frustrated man. As someone who was extremely reticent about my H, there were no friends and even family to rely on. Perhaps, if I had been religious I could have counted upon support from other followers, or indeed even the support of God. But I was indifferent and probably even a bit hostile to all religions. There's too much suffering and conflict caused by religion. Furthermore, many so called religious people I had met were not just ignorant, but often arrogant and hypocritical. But, I had been ambivalent about God ever since I was a young child. My rationality wants to dismiss the supernatural, but during my many moments of real physical hurt due to my H, I frequently crave it too. It was a natural and pragmatic desire for hope and best wishes that my condition would improve quickly, rather than a spiritual craving. Therefore, I remained an agnostic, rather than an atheist which I probably would have become had I been lucky enough to live a healthy life. But being an agnostic was not a comfortable position either. It conflicted with my scientific rationality for real provable evidence. Most damming of all, it conflicted with my personal anguish about a God who could sanction so much suffering in the world – including my own.

In February 1993, I resolved these conflicts in one of the most amazing time of my life. As mentioned numerously already, prior to the above period, I spent many days languishing on my bed trying to recover from the frequent bleeds that had blighted much of my life. The anguish and frustrations were really getting to me then. There was a sense of desperation until in an unrecorded date in February 1993 I had a profound flash of insight which changed my life.

My scepticism and doubt of religions did not mean I wasn't interested in finding out more about them. In fact, during my frequent bed bound state around this period, I had the opportunity to learn about the teachings of a particular 'Master' who was first introduced to me via a Chinese friend, Liu Wai. He was an engineer at the famous Rutherford Appleton Laboratory and we got to know each other when we were residents at Rush Common House. He lent me a couple of the Master's video tapes which I watched during my periods of illness and solitude. I expected to get bored and disillusioned by watching this kind of programme because they usually involve a combination of traits which I despise; boring and over-serious sermons, all-righteous preachers, exclusivity of the superiority of their belief over others, and followers who seemed bedazzled and brainwashed. This Master was completely different. She was extremely FUNNY, making jokes all the time, including self-deprecating ones. Unlike many religious or faith teachers, she was highly educated and had a very interesting and cosmopolitan life before becoming a 'guru'. Her teachings were inclusive, rather than exclusive. Anyone of any existing faiths (or non-faiths) are welcome. She appeared to combine the best of religious teaching and humanitarian principles. The central concept of her teaching is that everyone who behaves morally can become enlightened and achieve the highest levels of Awakening (or Buddhahood). To achieve this, she teaches a medication technique and a set of simple and decent precepts to followers: 1) Refrain from taking the life of sentient beings – she encourages veganism, 2) Refrain from speaking what is not true, 3) Refrain from taking what is not yours, 4) Refrain from sexual misconduct, 5) Refrain from using intoxicants. Well, that's all so reasonable and aspirational, despite not following all the precepts myself.

Besides being funny, I didn't mind watching a few more of her entertaining videos since they were quite informative too, reminding and teaching me some of the positives of different religions. By the way, the Master isn't 'perfect' in the usual sense of the word; in her videos, she demonstrates unremitting high ego with no pretence of false modesty! Also, just like a popular celebrity with a powerful and charismatic personality, some of her followers can be too devotional for their own good. Personally, I have never seriously followed her, but am grateful that I did watch these videos because indirectly her teachings inspired me to think more deeply on a spiritual level. Subsequently, I gained an insight which transformed my life. I will try to explain it in the following few paragraphs.

Of all the differences between the faiths, there appears a common theme in many great characters of religion and philosophy; they (e.g. Jesus, Buddha, Laozi, St Bernadette) all achieved their transcendent and sublime status (e.g. enlightenment, sainthood, some kind of spiritual high, so-called religious experience, self-realisation) through having endured prolonged suffering, and an intense devotion to a particular state of being, usually via meditation, praying or chants. Hence, there appears to be a link between suffering, focusing on internal devotion and transcendental status i.e. that

***Prolonged suffering + Focused internal devotion → Enlightenment.***

All of us know about suffering, although luckily fewer know of prolonged suffering. All of us also know about focused devotion, but this is usually external, such as devotion and love of a child or career or to the mundane and materialistic e.g. cars or possessions. Very few people devote time to their internal state of being i.e. by practising meditation or performing intense exercise that requires concentration. Therefore, as a matter of fact, very few people in the world can claim to have become enlightened. Hence, most readers may justifiably be asking at this point "What is enlightenment or self-realisation or spiritual awakening or nirvana or whatever similar words and phrases are used by other faiths?" Some might even ask what is this mumbo jumbo about enlightenment and self-realisation? Indeed, I held a similar disdainful sentiment up until the moment I realised a particularly powerful insight from my own experience of suffering. Furthermore, I was resentful about why suffering is required for enlightenment? If enlightenment is related to the divine, then is God sadistic in demanding suffering as a pre-requisite to enlightenment?

I will explain below the meaning of enlightenment from the context of the insight which I have gained.

This insight is very simple and powerful because it is based on rational and scientific thinking. It came to me during one of my miserable states of being debilitated where I surmised that I probably now qualified as having experienced 'prolonged suffering'. This suffering had made me miserable and angry of my state of being. Thus, I surmised then that if I meditated and developed more internal focus I might feel better or perhaps even experience enlightenment, whatever that meant! I think I tried that for a while, but nothing lasting came of it, except occasionally achieving some calmness during the meditations as my mind became empty instead of being filled with anxiety or negative thoughts. I realised that meditation, like studying is a particular kind of focusing of the mind, or training the mind to reach some particular mindfulness and peaceful state which is ultimately beneficial to the practitioner. Unlike ordinary studying where the goal is tangible (e.g. learning to play the piano, or trying to understand physics), the goals of meditation are not so tangible or readily achieved. I was sceptical if I could ever achieve these beneficial states.

Then one day in February 1993, I made my most revelatory insight in realising that when people suffer through ill health, their mind is naturally made to focus on it by the nature of its pain and attention that it needs. For example, when walking, most people will not be thinking about their knee, unless they had pain or injury in it. Hence, I realised that for almost 15 years of my life, ever since the traumatic bleeds at age 11, I have already devoted a significant part of my mind to being focused on my knee. Hence, just as a trained piano master (or a medication master) has parts of their brain overdeveloped/rewired as a result of daily training, that parts of my brain must also have been overdeveloped or become more attuned than the average person's brain because I have devoted so much time to thinking about it. Hence, I have actually already been experiencing deep *focused internal devotion* development for many hours and years; except I was not conscious of this fact, until now.

Eureka! It was like an incredible feeling of revelation. Instead of seeing the pain and debilitation of my right knee as a curse that had hurt and frustrated my life, I saw it as an opportunity to gain further *focused internal devotion* points! As soon as I realised that, I felt a tide of calmness, joy and elation wash over me. Perhaps, this is what it felt like to be enlightened. I also then suddenly understood the meaning of the phrase 'self-realisation' corresponding to the heightened awareness I felt when I worked out for myself something extremely profound – *that enlightenment can be explained as a natural cause-and-effect physiological change in the brain,* **without** *recourse to God or anything supernatural.*

The years of internal conflict about the existence of God disappeared instantly. No, I had not become an atheist either. On the contrary, the self-realisation moment had made me completely at peace with myself. The desire to know whether God existed or not just did not bother me anymore, intellectually, or spiritually. From that moment on, I felt I had become Enlightened.

The incredible part of this experience was that the blissful feeling was not transient. I was consciously aware of this state of euphoria[101] for the best part of at least six months. Honestly, I had this incredible 'high' without being on drugs of any kind! There was another aspect to my enlightenment experience that was truly marvellous; straight after the self-realisation moment, the cause of so much of my grief – the frequent and debilitating bleeds in my right knee were no more. Subsequently, there were far fewer bleeds everywhere and even when there was a bleed, it was much more minor. The enlightenment had literally liberated me. Certainly my mind was at complete ease when I did get a bleed, because I associated it with enabling me to get more enlightenment 'points', and perhaps even achieving a higher level of awakening. This wonderful stabilisation of my medical problem was like a deliberate response to my insight; you can't qualify for enlightenment points that easily now that you know the answer! For the experience of better health, that's fine for me too.

Of course, the reduction in bleeds could be just coincidental – it is known that the vulnerable inflamed synovial tissues in the knee could eventually burn themselves out. It is also accepted knowledge that a positive state of the mind can often affect one's physical wellbeing. But it really doesn't matter; the incredibly uplifting and special feelings that encompassed me from the moment of self-realisation are gratifying enough.

There were other good changes after my incredible experience; I became less self-focused and craved for opportunity to help other people in need. Unfortunately, there was one person that began to require my assistance at an increasingly frequent rate. Brian, my office colleague and friend had started to have more frequent relapses involving weakness of his formerly fit body[102]. Often, I noticed he would drag his leg as he walked. Like me, he was single and a very proud person, so I knew he would not ask for help unless he had no choice. Without prompting, I helped him out whenever I could, feeling useful but deeply anxious of his condition. Eventually, after many months of uncertainty, Brian was diagnosed with multiple sclerosis (MS), a degenerative and cruel brain disease which often robs the life of young people in their prime. I knew what it felt like to suffer at one's prime, but at least I could also see the light. During the early years, I often witnessed his severe debilitation as a result of his MS flaring up, followed by rapid recovery with steroid treatments. But with each subsequent treatment, it became noticeably less effective. I felt really worried and sad for him, which was particularly ironic because my own condition had finally become tolerably calm.

Finally, I wish to explain further how a 'self-realisation' moment led to a state of total peace on the question of God's existence. That peace corresponds to a sudden transformative absence of desire or need to seek the answer. As a thinking and curious person, this outlook is contrary to my nature, hence whilst writing this part of the book, I have been analysing further my feelings and experience more deeply. First, my newfound indifference to the question "does God exist?" was really about "Is there an external God being?" Hence, my total indifference to this question may well be because the concept of an "external God being" is too definitive and specific. I have, following enlightenment, experienced thoughts and feelings that simply entail inclusivity, in particular that individual self-consciousness is part of the whole (I could not honestly define precisely my meaning of 'whole' here, except to say that it corresponds to something much greater, e.g. whole is the world, or of everything and everyone that is connected to the self, or perhaps even everything; as in the whole universe?). In this state of existence and belief, the self and the whole are all somehow interconnected. There is thus a pervasive power that binds us all together. Some people may call this God! As a physicist, one might even ascribe God to the strange and interconnecting properties of the observable reality i.e. as observed in experiments which verified the entanglement property of quantum theory. And if I have to give a definition of God, or to define it, then I will choose to say "God is simply an energy that pervades everywhere in everything, or nowhere in nothing".

Anyhow, whatever God is, since my own enlightenment experience, the angst about who God is, or whether it exists or not just doesn't matter anymore.

## China's redemption

Following the Tiananmen Square crackdown in June 4 1989, Eastern Europe and the Soviet Union underwent their own momentous revolutions. Within a matter of months in late 1989, communism in Eastern Europe was shattered, while the Soviet Union was disintegrating under Gorbachev's disastrous but well intentioned reforms of Perestroika and Glasnost. After almost two years of political and nationalistic turmoil, even the mighty Soviet Union was liquidated in December 1991. The West was jubilant and triumphant. Could China be the next to fall?

At the beginning of 1992, China was still under international ostracization and a pariah nation. Its economy was stagnating from sanctions and factional politics which were increasingly veering back to hardline communism. The progressive reformers had been in retreat since the turmoil of June 4th, and leftist hardliners were becoming more powerful and defensive in the aftermath of the dissolution of the communist countries in Eastern Europe. Once again, it looked as though China just could not modernise without making a mess of itself because of stupid political ideology. It was painful and very disappointing that my wish of a prosperous and advanced China may just not happen within my lifetime.

*"It does not matter whether a cat is black or white so long as it catches mice"*

The above quote was by Deng Xiaoping, the reformer whose reputation was severely damaged by the Tiananmen Square crackdown. He was still a reformer and great pragmatist through and through. China was on a knife edge between economic and social collapse, and retrenchment back to unproductive 'leftist' orthodoxy. In January and February 1992, Deng who was by then 87 years old made his legendary 'Southern tour' to Guangzhou, Shenzhen, Zhuhai and Shanghai – the pioneering provinces and cities which led China's economic reforms in the previous decades. I can still remember vividly reading about the southern tour news in the Culham laboratory's library during lunchtime and immediately felt the awe of a tremendously uplifting event by a giant historical figure, visionary and reformer. My esteem for Deng became even higher when I learned that the tour had been planned with secrecy and with great vigilance as he had no official power at this point. Before the trip, he had pretended he was visiting relatives to deflect attention from leftist adversaries; and he had sought and enlisted the support of top People's Liberation Army generals to ensure his last great push to 'reform and open up' was not scuppered by hostile and opposing elements in the party or military.

The 'Southern tour' was initially ignored in the state media as Deng called for the people to wholeheartedly push for greater reform and to open up the country again. He knew brave progressive reform, not retrenchment was the only way forward for China. After more than two years of economic stagnation and dangerous political impasse, the entrepreneurial and hardworking spirit of the Chinese people was re-ignited. Once again, economic dynamism returned with a vengeance; hardliners and their backward ideology were side-lined. The rest is history, as China marched forward to become the world's largest trading nation in 2013, and is set to be the world's largest economy by 2020[103].

In the aftermath of Tiananmen Square and the fall of the Soviet Union, many commentators in the West wished or predicted the downfall of Communist China too. Some despised the governing communist party in China because it was non-democratic and didn't uphold the same high standard of (internal) human rights as advanced Western societies. Personally, I am adamantly against the sudden forced change of any complex country as it would almost certainly leads to chaos and countless misery. While China had enjoyed more than two decades of continuous growth and progress, the 90's was an extremely violent period in the former Soviet Union, the Balkans, and in central Africa; literally millions of people were maimed, killed or died prematurely due to civil wars, the collapse of their economy and health service.

Another successful legacy of Deng (and Margaret Thatcher too) is the 1984 Sino-British Joint Declaration treaty for the return of Hong Kong to China in 1997. Deng realised that HK and mainland China were at a totally different level of economic, political and social development. To ensure HK people's confidence, Deng proposed "One country, two systems", where except in defence and sovereignty, HK would retain its system of governing, i.e. economic policy, civil institutions, currency, freedom of press, and even the English law system would remain unchanged for 50 years from the date of unification. It was a stroke of ingenious pragmatism which Deng was so deservedly famed for. Today, nearly two decades after the handover and despite dramatic protests from the 'Umbrella movement' in late 2014, Hong Kong has stayed prosperous and stable[104]. Much of this successful legacy can be attributed to Deng's foresight and decisiveness. Deng's wisdom was to realise that economic, social and political reform must be prioritised accordingly. Too much political change without a minimum economic and social foundation is reckless and almost always leads to blood and tears. I agreed with him, and history has shown him right.

Coincidentally, in the late spring of 1992, I returned to Hong Kong and China for the first time since I left HK in 1978. I visited Wenzhou, Guangzhou and passed through Shenzhen. It was amazing even at this early stage to witness the development and vibrancy of these places. Moreover, I was also pleasantly surprised by the sense of humour of the Wenzhou people, while the famously blunt Guangdong people were not as rude or blunt as I had remembered from the 70's! But progress has a price; I miss the old steam powered locomotive which had been replaced by boring diesel engines, and more seriously, the small clean river outside my dad's ancestral home had turned into a stream of disgusting sewage. Also sadly too; HK people's civic manners were still appalling, and I felt annoyed and ashamed by this.

# 7. RETURN TO GOD'S COUNTY

My work at Culham laboratory could have been really fulfilling; the goal was noble, the research was very interesting at times, and the people were great. The latter was particularly important for me because when I was debilitated with H, I could still work from my bed by catching up with research papers or doing calculations. As long as one is producing the results, nobody minds. However, the reality of doing full time research is it could get stale; especially if the ideas are not your own. What I did much of the time there was carrying out research based on the agenda and ideas of my bosses. This is actually to be expected when one is a new researcher in the field; it is often said that it can take several years of work and study in the field before one can make an original discovery. However it is also possible to make an interesting discovery even at an early stage, as I did on my very first task, when I discovered an additional and unexpected mode of plasma instability for a radio frequency heated plasma. Unfortunately, I didn't have the experience or confidence to explore it further, thus reducing the impact of the discovery. By the way, when I say 'discovery' it does not necessarily mean something very significant – in science, most discoveries by working scientists are tiny steps, or paraphrasing one of Isaac Newton's famous quotes; "… just like finding a shinier pebble on the sea shore and discovering

194 *(1995-2002)*

something interesting and different about it". For most people, turning over many tiny pebbles and occasionally uncovering some tiny gems is the best that can be expected in their career. After three years at Culham, I was still just churning over tiny and occasionally slightly bigger stones. I felt frustrated by my lack of creativity, and this led to me doubting my own ability. In fact, being the youngest and least experienced I did sometimes feel like a fraud to be in such an esteemed group, sharing the corridor with brilliant people like Bryan Taylor, Chris Bishop and my office partner Brian Harvey who so casually solved their problems, or contributed to new ideas. Sure, H made my career much harder than it could have been, but my other fault is that my mind wanders too much meaning I get bored working in any one subject area for too long. It was time to find a new position that interested me fundamentally. It would be nice to do a PhD in the foundations of quantum mechanics – a topic which I had the privilege to touch on during my undergraduate final year project.

## Take it easy!

In my third year of trying, I was finally offered a PhD position back in my old university at York with recently promoted Professor Jim Matthew. My previous attempts to do my preferred research topic were cruelly scuppered because funding just didn't materialise in that area[105]. Hence, I was forced to switch to a topic which had guaranteed funding in place. Jim was a world expert in theoretical spectroscopic physics, so I would be working in this area for my PhD. It was not my most desired subject area, but at least it was theoretical and had some interesting atomic and quantum physics involved in the interpretation of the spectroscopic spectrum. I felt lucky that he had accepted me, especially as he was my favourite lecturer all those years ago who had also left two distinct marks in my memory: he gave the eye opening lecture on dimensional analysis[106] in my very first university physics lesson, and he had the most lyrically pleasant and thunderous Aberdeenshire Scottish accent, made more celebrated by an occasional stuttering intrusion. Later, as we had our not so regular meetings, I learned to sit at a fair distance from him to prevent his thunderous voice damaging my hearing permanently. He also loved talking about the great Scottish physicist, James Clerk Maxwell; the father of electromagnetic theory and therefore, of modern communication technology. Jim was another of those all too common resplendent Scots who are frequently found at the top of many English institutions.

After almost 5 years, I left Culham, feeling excited for the future, but also sad to be leaving many nice people behind. I would especially miss Brian and worry for the uncertain future that MS wrought on his life. On the other hand, I was also relieved because my own experience of suffering had made me particularly attuned to his. It would be a harrowing and sorrowful period to witness the cruel decline of a wonderful friend. To this date, I regret I kept my own condition secret from him; not to burden him with my problems, but the act of unleashing my demon may also have helped unleashed his.

Other than my wish to switch subject and gain a PhD, there was another major motivation to leaving a good job, swapping a decent salary for a grant of just £6K per annum. I didn't mind about the financial change at all as money was becoming lower and lower priority in my life; good health and well-being had become my dominant motive, and treasures to sought for. I wanted more flexibility, less physical pressure, and to be more relaxed about life, while still doing useful work. No, I don't mean being lazy, but rather just being able to rest properly and not worry so much about not being able to go "into" work when I had a bleed. It had to be said though, my previous job had also been flexible with kind bosses who never complained (even though, I think none of them knew about my H). There was nevertheless always some guilt and discomfort when I had to take sick leave[107], notwithstanding that I had often pushed myself back in the office while just barely recovered. Hence, doing a PhD was a great privilege to lessen the physical pressure of a conventional job.

A PhD is a long term project lasting between 3 to 4 years, so I thought there was plenty of time to enjoy the full social life of the University. Indeed, I didn't work hard in my first year, preferring to re-live the best of a student lifestyle, somewhat suppressed previously; making more friends, socialise more and having fun. I went to more parties and chill out a lot more. Being wiser, and more confident, I also got to make many more friends of various nationalities and age groups. I really felt truly cosmopolitan and socially more appreciative than many of my contemporaries. For example, I found the bulk of native British students tended to make little effort to befriend overseas students, being rather parochial, and because they were so naturally at ease in their familiar environment, they likely felt they needn't bother. Similarly, but for the opposite reason many overseas students (Chinese and non-Chinese) also tended to socialise with each other almost exclusively.

My time in Culham was barren years for love, so I hoped to change that too. I wasn't shy at meeting and talking to girls, unlike ironically many of my contemporary English friends in the physics department. Most of them seemed to prefer the company of messing about with computers, installing new Linux system kernels, drinking and getting pissed. Without alcohol, frankly I think most English men are really rather shy and reserved; actually much like me, except I realised this and tried to make an effort. Similarly, I found most English gals at university also quite offish and parochial, in contrast to more mature English (especially Yorkshire) women who are generally very friendly and open. On the other hand, overseas female students were far friendlier; Chinese, French, German, Japanese, Korean, Italian, Spanish, Thai and Turkish girls have all been great company. I wanted to be more than just friends with some of them, but there were always hidden barriers between us; some were cultural and personal, but 95% were barriers which I erected to protect the secrecy of my H. Often I had to call off dates or avoid them altogether because of a bleed or the discomfort of trying to walk normally during a date. And when a date was going well, I inhibited my attraction and desire for the girl because I feared infecting her with HCV – I don't mean sex, which would actually be quite safe because safe sex is expected. It is the more innocent and intimate act of kissing that worried me. In fact, I found my HCV status to be a greater hindrance to my desire to find love, than my H. I may not like revealing or talking about my H, but the former condition is a serious health issue for any potential partner. Although medical evidence has shown that HCV is extremely unlikely to be passed on by kissing; during dates I remained aloof and acted more like a great gay friend, showing the girls a great time, but avoiding the physical contact that is so often required to start a relationship. My strategy was that I wished some of the dates would fall for me first so that I could reveal to them my conditions at the right moment. Unfortunately, such girls are rare. Sometimes, I feel a bit bitter that the expression "good guys don't get the girl" rings so true, especially when some asshole character gets the girl instead. But, don't feel sorry for me though; I never felt bored with being single – my situation was in many ways better than many of my 'normal' friends and acquaintances around the physics department who apparently didn't bother to talk to any girls at all.

## Yorkshireman and the motorbike

When I was living in Abingdon, my landlord who was a softly spoken civil engineer and about my age had a big motorbike; a Honda CBR1000 grand sport tourer. He commuted on it every day, firing up the raucous engine roaring aggressively when started, while its idling could be felt from my bedroom as the Honda engine transmitted a powerful but subtle thunderous vibes through the ground. While waiting for the engine to warm up, he would transform his slight frame into a macho biker; leather jacket, boots, and gloves endowed his body with a cool helmet covering his head. As he mounted the big bike and rode off into the distance, the roar of the bike, the image of his leather clad body striding and controlling that monster machine was just so bloody cool! Until then, motorbikes were just a curiosity; it never crossed my mind to ride one; they were just too damn dangerous, with accident and injury being an inevitable fate of most riders. I had a sports car then, a limited production Mazda 4x4 Turbo, which looked like a slightly souped up version of the rather dull 323 model which it was based on. It was Mazda's attempt to sex up its range by making a car which could be entered in world rally competitions, competing against the likes of the legendary Audi Quattro, Lancia Integrale, Ford Cosworth and Toyota Celica. Surprisingly, despite only having a 1.6 litre engine, the Mazda was moderately successful in the world rallying circuits. I loved the car as its understated look bellied its brutish performance which often gave me the pleasure of burning off the odd Ford XR3i and Golf GTI boy racers. The four wheel drive gave the car that extra stability in the wet and in bends. It was fun, safe and fast, but compared to the landlord's bike, it was nothing. I started to wish I could ride a motorbike. I couldn't help but visit the local Honda dealer – admiring and drooling over the exhilarating and legendary Honda Fireblade. The model with the famous orange tiger stripes was just stunningly beautiful. But being a severe H, quite badly disabled in my right knee and with a weakened and arthritic left elbow, the thought of me ever getting to ride a motorbike was just mad and suicidal.

When I returned to York to study, the freedom afforded to me reignited my dream of becoming a biker. I am not exactly sure what drove me to this dangerous desire – I suppose it was me wanting to reaffirm my normalness, not wanting to let my H deter me further from a normal desire that normal people often take for granted (actually, I'll be honest with you, I rather liked the posing value, and wanted to be more attractive to the girls too). The desire turned to reality when I enrolled in the local motorcycle training school. I suppose it was a bit mad of me with my condition to even contemplate such a dangerous activity. But, I was not reckless or suicidal; my thought was that many people ride safely and don't get hurt. Hence, I would be part of that group, by riding safely and being extra careful. Being an experienced driver, I already had the road craft and awareness of the dangers of driving; riding a motorbike was therefore just a case of learning to handle a two-wheel vehicle. Indeed, before I had my first lesson, I'd been commuting with a pedal bike, perfecting my balancing skills, and occasionally pushing it round bends a bit faster to gain extra confidence. Also, I bought and read several books about the theory of riding a motorcycle safely. Furthermore, I bought high quality motorcycle clothing with high specs armours that protected the knees, elbows and shoulders. My hands and feet were protected by proper motorcycle gloves and boots respectively, and of course most importantly, my head was endowed with the best of Japanese full face helmets; the Arai Quantum, as worn by five times MotoGP world champion Mick Doohan.

I had entered the world of motorcycling; a close knit community of enthusiasts sharing the joy of two wheels. I felt great to have survived my first day of compulsory basic training. To be frank, I was not even sure that I could actually handle a motorbike, which unlike a pedal bike is significantly heavier. The weight can be really intimidating at low speed as if one loses the balance, the bike will topple over. My left arm was significantly weakened from years of bleed damages in the elbow joint – and similarly my right leg couldn't really handle much stress either, as the old knee was in a pretty dire state with limited bending range! However, I predicted it wasn't going to be a problem, because fortunately, the default and professional style of supporting a motorbike while stationary is via the rider's left leg (my good one), while the right leg always remains on the foot pedal. Hence, as long as I had good balance at low speed, riding a motorbike should be no more strenuous than riding a bike.

After years of mostly mingling with scientists, academics and students, it was particularly nice to mingle with non-academics, appreciating their more down to earth sense of humour and interests. There were many anecdotes about learning to ride; there were some horror stories, like a learner losing control on her first day, injured and never dare to get on a bike again. Another story was more delightful as Catherine Zeta Jones was taught by one of the trainers in the school. Presumably, I guessed it may have been for her role in that ridiculously 'chirpy and perfect' TV series "Darling Buds of May". The trick of handling even a light learner bike was challenging and delicate. I was in awe at the trainers with their experience and casual ease of handling their big bikes.

Some of them would often show off their skills with fun antics such as pulling a wheelie or touching their knees on the ground. Hilariously, I was most impressed by trainer Tracy, who somehow managed to light and smokes a cigarette whilst doing 40 mph on her bike.

After several months of training, I passed the test at my second attempt. I felt a tinge of triumph for this mischievous achievement. It wasn't without mishap though; during training for emergency braking, I grabbed the front brake too hard, the tyres locked up which threw me off the bike. One of the foot pegs had digged into my boot while my body and hands scraped on the ground for what seems like a 10 feet slide. Fortunately, being fitted with protective gears, I got up without any injuries, except hurt to my pride and the need to replace my trousers and gloves. It makes me angry when I see motorcyclists without proper protective gear, as incidents like this show that even a relatively slow accident can lead to serious injury; had I not worn boots, the bike's foot peg would have almost certainly broke my leg. Having passed the test, it was tempting to get a nice big bike as small bikes are not cool. I resisted this temptation and compromised with a stylish import, a Suzuki 400 bandit; a quasi-sporty retro-looking bike, which was relatively light and easy to handle. Nervously, I took it out for a first spin, getting a better feel of the bike and adding more hours to my experience. Embarrassingly, I had a slow speed accident on my very first ride. As I attempted to turn the bike round in a cul-de-sac, it tilted over a little more than I wanted. The heavy motorbike just toppled over! Bugger, all this happened in front of some other nearby bikers who then came to help me. Instead of looking cool, I looked more like a twerp. The incident didn't put me off though, gradually and over time I increased the range and challenge of my rides, first pottering along simple roads, and then venturing further afield into the surrounding North Yorkshire countryside, blessed with great scenery and twisty roads. I was loving it. Even slow speeds around town in the grid-locked city centre became fun. Oh God, I love filtering through traffics. This act alone makes me appreciate the meaning of total (road) freedom! All was going well, and after less than a year, I upgraded to a Kawasaki ZZR600, a powerful looking sports tourer bike with a broad seat and a cool twin exhaust system.

At this time, I joined a local motorcycling club – the York Advanced Motorcyclists (YAM), a group of local enthusiasts dedicated to the joy of biking while teaching the acts of safe riding. The group met every Saturday morning where experienced and qualified volunteers would tutor new members on how to ride effectively and safely. The tuition is given free of charge, and was massively fun as the tutor and students explored the roads of North Yorkshire – the biggest county in England with arguably some of the best biking roads in the country. The tutor observed the students' rides and they were then given feedback to improve or correct their road craft. After my very first lesson, I realised I had been riding 'wrong' all the time in my previous year by failing to move my bike proactively to maximise visibility of the road ahead. That is, when one is coming up to a right (left) hand bend, the rider would move from the standard middle of the road position to the nearside (offside) of the road. This enables earlier observation of the road ahead thus allowing the rider to take the road's optimal path more safely. This simple realisation almost immediately enabled me to ride more smoothly, faster and more safely. For some reason[108], this simple technique wasn't taught during the learner lessons. It was a reminder that progress through self-learning, without teachers or peers to learn from, is usually extremely limited. Also, trying to keep up with other more skilful and faster riders forced me closer to my own limit. Thus the limit of my capability was steadily increased[109] with every new ride.

One of my biggest joys of riding in the club was getting to visit many beautiful places in Yorkshire, and sometimes beyond. Ironically, I first visited two famous physics places as part of the riding tours organised by the club. After riding through a spectacularly series of twisty and undulating roads across the Yorkshire Dales, and through to Lake District, we saw Sellafield nuclear power station in the distance – the world's first commercial nuclear power plant on the west coast of England. The other physics related visit was a mysterious ride across the Pennines to Cheshire, near Manchester. When we arrived, I was elated to discover that the destination was Jodrell Bank, the site of the world's first radio astronomical observatory. I remember vividly reading about this place in my childhood as I imagined the vastness of space, and thought about Einstein. Never would I have imagined that one day I would visit this place with a bunch of bikers, in my leather clad gear, having arrived in my own sports bike! There were opportunities to visit Scotland and even continental rides, but I never went on these because I couldn't trust my body would last the long journeys without mishap. However, the local rides were good enough for me. I will never forget the many thoroughly blissful rides through the spectacular countryside, quaint towns and villages of the Yorkshire dales, the Yorkshire moors and the wonderful seaside towns dotted along the Yorkshire coast. Oh heaven, I felt so privileged to have this experience living in beautiful York, with great Yorkshire people. No wonder Yorkshire is known as God's county.

There was another intangible feel good factor of being in a club –the camaraderie and friendship with another group of people which I wouldn't have meet otherwise. At University, I got to make friends with people from all parts of the country, and of the world. This was extremely enlightening and educational, but ironically that environment, as diverse as it is, it is also ironically quite insular; in an ivory tower kind of way as some people would say. In the YAM club, I got to know a more colloquial but nevertheless diverse bunch of Yorkshire men and (a few) women. There were salespeople, a painter, a teacher, a retired executive, a courier who was an ex-police rider, a decorator, business owners, builders, retired professionals, a masonry craftsman, and others. I rather liked these friendly Yorkshire folks with their down to Earth attitude, sense of humour, and affable dialect. Interestingly, although not surprisingly being in a minority hobby in relatively rural North Yorkshire, I was, as much as I knew, the only ethnic minority in the club! Hopefully, my presence may have contributed some education to them too; that not all Chinese work in restaurants or takeaways!

I gave up motorcycling in 2008 because the knee finally deserved its long overdue repair[110]. Although, there were a few close shaves, I rode for more than 10 years without any major mishaps. To be frank, I would not recommend anyone with severe H to contemplate motorcycling. Looking back, one needs luck and precaution to avoid potentially dangerous misdeeds. I certainly have a cautious side in me; in one long ride with some crazily fast riders, one of the guys, Barry, said to me, "Have you got a safety switch in-built in your head?" I proudly said, "Yes". I knew too much about pain and suffering to risk an accident. Unfortunately for Barry, but not entirely surprising he broke his collarbone a few months later in a motorbike accident. I think the people who knew me in the club would be surprised to learn their only ethnic minority member, was also a cripple, and has a dodgy medical condition. Hey, I am still alive, and had a tremendous time during my membership. Thank you very much, YAM. I miss you all, and maybe one day I will return!

## A dot-com adventure

The PhD or doctoral thesis is the king of all degrees, the highest academic honour which a University can bestow on a student. Although, I had in effect postdoctoral research experience before doing my PhD, I still imagined that only very clever people, full of original and creative ideas could complete a PhD. Well, I'm not going to pretend that I'm not reasonably smart, and according to my friends, I'm certainly a bit different and eccentric; a mark of genius perhaps? But the reality was as much as I do think and behave differently to other people, I hadn't been particularly outstanding at my research job, or produced any great ideas either. What I seemed to be quite good at was constantly dreaming up inventions that weren't related to my work or study. Sometimes, I like to blame H because it does so frequently disrupt my life, but if I were to be honest, my character lacks the one dimensional focus of most conventional and successful scientists whose dedication to one main scientific discipline throughout their lives simply bores me. So I found myself rather adrift and lay back in the first two years of my PhD, helped also by the fact that my supervisor Jim was also very laissez-faire too. However, there was no escaping the hard work required to succeed. In the end, I did finish the thesis, reading and writing intensely in the last 12 months right up to the final hour of a four year deadline. It was a really memorable time too, as a select group of PhD students were frequently around the department most evenings and early mornings – working hard and socialising hard too; in between work, we had great fun chatting, playing video games, discovering music and interesting stuff from the then nascent and increasingly expansive internet, eating, drinking, and some of us smoking during breaks in all those eerily odd hours of the day. There was a wonderful sense of camaraderie as we all laboured with the 'fun' of writing up our thesis. By the way, here's a synopsis of how to get a PhD and what it's all about. A PhD is essentially a substantive piece of original research into a particular subject matter. The emphasis here is on 'original' and 'substantive'; what you do must be new and not just a rehash of other peoples' work. I used to be quite daunted by these two key words. But actually being original does not mean you have to come up with extraordinary insight. For most people, just approaching an existing problem in a different way, or repeating the same experiment but investigating a new compound is enough originality. As for substantive, using the same example, if you had investigated sufficient different compounds then that is in enough work to make it substantive. Of course, you will have to interpret the results correctly which is where you need to have established a sufficiently critical mind and knowledge of the subject. Hence, a PhD starts with an area of research which you, or more likely your supervisor, wishes to study. Sometimes, you or your supervisor may have a very clear idea early on in your research, so you can immediately start with a thorough literature review. It is crucial you know all the prior knowledge so you don't waste your time doing experiments or researching specific subjects which have already been dealt with before. This sounds easier than the reality of trying to understand the prior work. Often there is further background or jargon to comprehend, and in the case of theoretical physics, very likely there would be many difficult equations to work through. The literary review can be really exhausting to the mind, but over time, and partly subconsciously, the knowledge and wisdom in these books and papers is distilled inside you. Consequently, the mysterious way of the mind will enable you to filter and seek out a new areas to investigate. Thus, to come up with new stuff, one often needs to know the old. But sometimes one may find that there is no existing prior literature in a particular field, or where it does exist the explanations or theory proposed is not convincing, thus leaving room for some alternative (hence original) explanation. So getting a PhD doesn't really require an exceptional mind, but it does need a persistent and hardworking attitude to get through it successfully. It is often said that academics don't make good salesmen. This may be true in the business world were they to change job. However, one big lesson I learned is that the most successful academics are, in fact, some of the best salesmen in the world – that is they see even small ideas as important and deserving of publication and public funding. In fact, amongst the academic and research circles, we scientists often recognise that the fastest career climbing researcher is often also the biggest self-promoter (or a bull-shitter as in the case of a few people I knew) too. I have met a number of these people in my career. I don't like them because from personal experience, these people are also frequently egocentric and selfish about helping new researchers. For me, the biggest lesson I can pass on from doing the PhD is the importance of learning the attitude that what you may think of as a small or an insignificant idea may actually be important, and hence it is worth developing the confidence not to undersell yourself or the idea. Certainly this is the attitude which you must take with you during the viva for the PhD thesis. Not arrogance but assuredness that what you have done has made a significant contribution to the sum of human knowledge and that you are deserving of getting the Doctor title. After just over four years of work, I underwent almost three hours of an oral examination by two examiners, one internal and one external.

The job of the examiners is to analyse your work, pose queries about it, and critique it if they disagree or don't understand part of the work. I would have to explain or defend it. The explanation parts were relatively straightforward as I had combed through my thesis, making notes next to parts I thought the examiners might query. You are allowed to bring along an annotated thesis with you, which was advantageous to me because my 'physics' memory is not great. However, the critique parts are unknowns and how well you defend depends on your understanding, knowledge, argument and confidence (yes, ability to bullshit is useful here too!). This is potentially the most treacherous part of the viva – you could fail it if your argument or understanding falls to pieces, or you could be very unlucky because one or both of the examiners turn out to be a nasty person who just want to do you in, or are jealous, or too proud to accept your exceptional discoveries. There were many questions in my viva. At one stage, I couldn't answer some basic questions which was really worrying. However, I also recall answering well to some of the trickier ones. I was told to leave the room for their deliberation. It may only have been about 10 minutes outside in the corridor, but felt like ages! When I was invited back, I saw the examiners' faces, inscrutable for the moment, then with their hands extending towards me, and their faces lit up with smiles, I heard the words, "Congratulations, you have passed….". Suddenly, I literally felt like a big weight had dropped off my shoulders. It was a WONDERFUL feeling. That evening, I had the best sleep of my life – except perhaps excluding the time when I was a carefree baby!

There is one more interesting anecdote about my thesis; at the beginning of my viva, the external examiner Professor Weightman of Liverpool University remarked to me that my thesis had been a delight to read, praising my writing style.

I felt really good about that as even then I knew I wanted to be a writer, so he was the very first person to pass positive judgement on my writing style. For those of you who might be curious about my thesis, the title is "Theoretical studies of quasi-atomic excitations by electron and ion impact". I learned a lot, but to be honest, the subject didn't light my fire.

The beginning of the millennium was a truly exciting time in the world of business. The dotcom boom was at full steam and I had got my PhD out of the way. Like many people who have just completed their PhDs, the thought of continuing in the same field didn't appeal to me. Besides, this dotcom boom was infectious; we were in a revolutionary time in the business world. Companies were starting to get online, and e-commerce was just beginning. Ridiculous stories of youngsters receiving multi-million pound investments for nothing more than an online idea made the thought of starting up a business the most exciting thing anyone can do. I can't quite recall the exact moment, but sometime shortly after my viva, I became absolutely determined to be part of this revolution. With no prior experience of management or building a business, I imagined the creation of a 'Chinese portal' for the Chinese community in Europe. After following China for many years, I was certain that the rise of China's middle class was inevitable and very soon there would be tremendous business opportunities between China and Europe. But I had neither technical skills nor the finance to create my dream. I needed to form a team who would share my vision, and maybe I could then raise the apparently easy capital that everyone seemed to be getting. For the first time in my life, I started a company named Fusion 2020 Ltd., inspired by my past occupation, but also looking forward to the future of a joined up world. But already in early 2000, the dotcom boom had suddenly become the infamous dotcom bust – there was no easy investment money.

I didn't give up though; such was the momentum and desire which had been building up. I had already recruited Kevin, a recent York graduate from Hong Kong. He was much better connected to Chinese culture and language than me. We supported each other as I sought to recruit the other missing elements – technical guys who could create websites. I persuaded my less adventurous younger brother to help out too, often causing a lot of friction between us as I pushed him to help me with the graphics and website design. Eventually, after months of uncertainty I persuaded two of my contemporary physics PhD friends to join in my dream; they had the technical skills to create almost any website. My team was complete, or so I thought until my then best friend John, an ex-soldier, jack of all trades, turned biologist and lover of numerous women pestered me to join the company too. As he had some previous business experience, I let him join the company as finance director. Hence, we now had a sales manager in Kevin, technical director Geoff, software developer Steve, and me as the managing director. We were now all full time risking our time and money to build a dream. At last, we were altogether in an office, riding out our own dotcom dream, except we had to do it by our own bootstraps rather than investors' money. In order to ride out the dotcom bust, we made the decision to earn some income by making websites for local businesses. Now, I found myself doing things which I used to hate, cold calling businesses trying to sell to total strangers. It felt so alien for a person with a scientific background, but being the leader with the responsibility of the team, I found the courage to do it. Surprisingly, the team managed to build quite a number of websites in a short period. However, we were barely keeping up with the costs as we were obviously undercharging. Our PhD experience had taught us about selling our research contributions, but when it came to commercial valuation of our work, we were scientists at heart – money was a dirty word. The company could have benefited greatly had there been an experienced mentor to guide us then. I also regretted having recruited my dear friend John. Although he was supportive at times, with hindsight, the overwhelming influence of John was a big distraction as he became bossy with too many ideas, and his all too frequent ill-tempered style sowing discord and disharmony in the team. My original vision of building a great internet portal was quickly subsumed by these distractions. Despite not exactly bustling with work, I appeared to got myself tied down by low priority administration while lack of finance and experience curtailed our marketing and sales strategy. Although I felt I could pat myself on the back for putting the team together, managing the team within an office environment was an entirely different situation. In the beginning, there were squabbles about who sat where, or who was more senior as John started flexing his 'Director' status. The open-plan office environment felt painfully awkward at times. As the whole team were really just volunteers rather than salaried employees, the command structure was meaningless; my power as the CEO of the company was purely via motivational talk and persuasion.

After more than half a year of a pretty intense emotional roller coaster ride for all of us, it was becoming clear that the company was not sustainable. By July 2001, the team had started to disintegrate with resignations coming in on a weekly basis. My last hope was to do a pitch at a business angels event in Hull. It was a slim chance really because most investors would rather punch themselves than entertaining any more dotcom businesses. As expected, there was no last minute saviour. However, there was one portly guy with a frizzy balding hair who showed an interest in the proposal. He was Robert, an experienced Leeds based internet entrepreneur who had already cofounded several internet based companies. The dying company had received a reprieve as we negotiated a deal to create the portal of my vision. But there was no direct cash investment, Rob would add his expertise and find someone to create the website. The old team at Fusion2020 disbanded anyway. It was sad, but everyone was relieved. The company had no debt, and all our clients were looked after properly until the end of their contracts. Kevin returned to Hong Kong, Geoff and Steve both found professional jobs befitting of their computer skills, and John continued as a biologist. I was relieved too, but my entrepreneurial journey had not yet ended.

The negotiation and process of enabling my vision now solely lay with Rob and me. As the process was moving slowly, I admitted to Rob that I too needed to find a full time job to replenish my savings. To be frank, at this stage I did not have high expectations that my vision would materialise soon. There was too much of the same slow improvised website building process I had experienced earlier with Fusion2020 when I begged my brother to create the company's first website. Rob was doing the same to his technical partners who were already busy with their own businesses. Well at least the dream was still dangling and Rob was sorting out the company accounts, saving me the cost of accountant fees.

Not long before I met Rob, my company was contacted by an attractive Chinese woman called Suzy. This was the first time that I encountered a mainland Chinese woman who wasn't a student or working in the catering trade. She was just a couple of years older than me but seemed very sophisticated, with a strong business background. She had an MBA and had been working for a UK manufacturing company until recently. Hence, she was looking for new business opportunities which, through the "First Tuesday" networking club, led to our first meeting. When she arrived at my office, she must have been quite impressed with me as there were three other full time staff all busily working on their computers. Knowing the imminent fate of my company, there was not much I could do to help her, pretending that my company was too focused with current work to cooperate. She left impressed, not realising how dire we really were. Then a month later, she called back. That time, I told her my situation and tried to help her by introducing her to Rob. But my kindness led to the final fatal demise of my portal vision. Suzy turned out to be extremely influential to Rob. I have no doubt that he had become besotted with her, as she turned up her full Chinese female charm. Behind my back, she persuaded Rob to adopt to her e-business idea; selling high quality Chinese art and clothing products targeted at the UK and European consumers. Fusion2020 was finally abandoned, while a new e-commerce website emerged. Although I became a cofounder and a small shareholder of this company, I felt betrayed by Suzy. Later she explained her innocence. But I have never forgotten this act of betrayal and inconsideration. It taught me a lesson that business people can be absolutely ruthless, innocent or not. I remained friends with Rob and Suzy[111], occasionally offering my words of wisdom to them. I was tired of business and decided to return to science again. A few months after the demise of Fusion2020, I managed to find a proper job again, as a postdoctoral researcher in the Electronics department at my old alma mater.

On 1st April 2001, while still frantically running Fusion2020, I received a phone call from my ex-Culham colleague Colin. His usual cheery and lyrical Scottish voice was sombre – Brian had died. Although, not unexpected, tears flooded my eyes from spontaneously thoughts of his latter tragic life and the happy memories of our friendship. The world had lost a brilliant mind, and a pure soul. This is the final part of the eulogy I wrote shortly after his death – *"I prefer to remember Brian's loud laughs, sharp wit and his (often sardonic) sense of humour. Perhaps this is why he chose to leave on April Fool's day[112]"*.

## When demons fight

On this same day, a foolish maneuver by a pilot caused a major international crisis between China and the US. A top secret US spy plane had collided with a Chinese jet fighter near Hainan Island in Chinese territorial air-space. The Chinese fighter pilot Wang Wei was killed while the US EP-3 spy plane, packed full to the brim with top secret surveillance equipment and data was forced to make an emergency landing on China's Hainan Island. The 24 US crew members did their best to destroy the on-board sensitive equipment and data, while Chinese soldiers attempted to forced them out of the stricken but securely locked plane. The crew were eventually detained 15 minutes after landing. China was furious, expecting America to explain its actions and show remorse. But no, instead America retorted even more angrily demanding immediate release of their crew and threatens China with serious actions. I still recall vividly the scathingly aggressive tone of Dick Cheney and Donald Rumsfeld making dangerous threatening warnings to China, while creating the false impression that the crews had become hostages, and with the use of misinformation insinuated the Chinese pilot guilty of causing the crash.

Over a pub lunch in the Rose and Crown pub near the company's office, I recall discussing this incident with Geoff and other physics graduate friends. I could hear them concurring with the American claims blaming the Chinese pilot as the culprit of the crash. Their arguments were based on the US media release of 1) previous footage of pilot Wang apparently flying recklessly close to the EP-3 aircraft, and 2) that the heavy EP-3 propeller plane was a bulky vehicle unlikely to be the cause of collision with a fast, agile jet fighter. As a physicist, as were most of them, I was quite shocked at how easy even intelligent and educated people were taken in by such simple misinformation. I told them, if you look carefully at the film footage, there was no evidence that pilot Wang had made a crazy mid-air close approach to the EP-3; it looked more like zoomed in footage, showing the Chinese pilot cockpit apparently getting bigger in the frame. The second point, I just rebutted by saying that was like saying that a crash between a lorry and a Ferrari on a motorway must always be the fault of the sport car driver. I am not sure if I got my points across, but it was obvious that most of my Anglo-Saxon English friends were on the side of the American government. There was another factor in their biased support, China had been demonised as an aggressive Commie-baddy who were holding the crew as hostages and maltreating them. As someone who was very aware of China's foreign relation's record, I thought this aggressive PR against China was crude and very harmful to China-US relations. I knew and predicted that the crew would be treated well, and with reasonable diplomacy they would be going home soon even though under international law, China had the right to detain the crew as prisoners and prosecute them in court for spying. Hence, I came out defending China's case as the US was caught in the wrong, and also because I had become infuriated and worried by an increasingly hubristic and aggressive neoconservative (or neocon) agenda of the recently elected Bush administration. Then, out of the blue, one of my most trusted friends and colleagues said something to me which hit me hard, *"You are in our country now!"*

I felt shook and hurt by this indirect statement of exclusion. It was as though that because I am ethnic Chinese, my defence of China implied somehow that I was disloyal to my home country Britain, a sometimes too loyal ally of the United States. I ignored the statement as I was too tired and shock to open a debate on my identity.

As for the outcome of the spy plane incident? After over a week of tense standoff, the US apologised to China. US Ambassador Joseph Pruher delivered a letter to foreign minister Tang Jiaxuan which expressed "very sorry" for the death of the Chinese pilot, and "We are very sorry the entering of China's airspace…". The crew were released and the whole incident ended peacefully and undramatically, as I had predicted would happen if there were no gung-ho reactions from the US. I must admit that I felt rather proud of being (ethnic) Chinese in this instance, because the aggressive bullying postures of the neocons had been humiliated by China and put in check by the pragmatic voices in the US administration.

The spy plane incident highlighted an increasingly anti-China sentiment in US politics over the previous decade. The nineties were an extraordinary turbulent decade following the demise of communism in Eastern Europe, the break-up of multi-ethnic states in the former Soviet Union and Yugoslavia, Chechen wars in the Caucasus, and genocide and civil wars in Rwanda and Congo. Even South Africa under the brave leadership of Mandela and de Klerk experienced numerous bloodbaths before the destruction of apartheid. These were extremely bloody events which many commentators in the West saw as inevitable, with some (usually armchair critics living a lovely life in their stable and peaceful country) even seeing these events as desirable; a necessary adjustment for the sake of a better democratic future. Amongst these tumultuous events, China had remained stable and made rapid economic and social progress due to the revolutionary reforms re-set in motion by Deng's legendary southern tour in 1992. Hong Kong was successfully returned to China, without all the doom and gloom predicted by some people. China and Hong Kong were the only major economies in Asia that weathered the 1997 Asian financial crisis. I was extremely pleased to see that China had been the bastion of peaceful reforms and economic management; instead of going through more chaos or revolution. My personal view is that so long as China follows a path of continuous reform based on sound government policies, this is better than the inevitable chaos which would result if it prematurely pursued the idealistic goal of achieving democracy at whatever the cost. I want to emphasise that I love Western democracies, but I do not believe democracy is the magical panacea of solving the ills of China and many other countries.

All was going well with China, except Taiwan. During the nineties and most of the noughties, Taiwan was governed by pro-independent separatist president Lee Teng Hui followed by an even more pro-independent provocateur and corrupt Chen Shui Bian[113]. The policies of the latter were designed to provoke China and create discourse between China and the US, while he profited from his extraordinary corruption spree in office. For the neocons, China was an anathema, a successful one-party authoritarian communist state, without democracy and poor human rights record which they increasingly saw as an enemy of freedom and potentially opposing of their ultra right wing agenda. Taiwan had become a major flashpoint which they could use to challenge China; by encouraging pro-independence forces in Taiwan, and agreeing to supply them with advanced weapons and vehicles. It was apparent to me that the Bush administration had started a containment and increasingly belligerent policy toward China. Most dangerously, they appeared to want to promote Taiwanese independence to challenge China. I felt increasingly exasperated with America's over-confidence that its new perceived hyper-power status could influence and win any causes and wars it saw fit. The demise of the Soviet Union was triumphantly celebrated in the West as good over evil, and it looked as though the US was desperate to create a new enemy. In particularly, I feared that a conflict between China and US, over Taiwan was almost inevitable and imminent.

On a sunny day in September, I woke in my parents' home in Manchester feeling somewhat uneasy. I had an unusual bad dream; I can't remember the details, perhaps it was something to do with the rollercoaster ride experience of the past eight months – surviving and eventually abandoning the dream of running a successful start-up. I started to look for conventional employment again. For a brief while, I considered investment banking as a possible temporary career to replenish my depleted savings. I surfed the internet and chanced on the US website of Morgan Stanley, a major investment bank with its HQ in the World Trade Center in New York. Delving deeper into the website, my mind hesitated about the wisdom of working in a big city. I didn't think much more as the website ceased functioning, and then followed by the internet slowed down to crawling speed. Frustratingly, I turned on the TV instead, and emerging from the flickers came the almost unbelievably surreal aftermath of a plane which had apparently crashed through the north tower of New York's World Trade Center. It was truly the most unbelievable and spell bounding live images ever broadcasted. Incredulously, that was just the beginning of horror on that infamous day. Under Osama bin Laden's command, Al-Qaeda committed the most deadly terrorist acts against the USA which brought down four airliners, totally destroyed the two towers of New York's most iconic buildings, and severely damaged part of the Pentagon. Almost 3000 civilians lost their lives on September 11 2001. With the exception of extremists and brainwashed Islamo-fascists, virtually everyone in the world was outraged and in sympathy with the American people. The initial US response was measured, to destroy Al-Qaeda and its leaders in the rugged terrain of Taliban controlled Afghanistan. After failing to kill or capture Osama and the Taliban leader by heavy aerial bombardment, the US and her allies invaded Afghanistan. It was a necessary and understandable action of a very angry and righteous nation. However, instead of focusing on the terrorist groups in Afghanistan and Pakistan, the neocon in the US initiated a much bigger agenda and vision; to alter the geopolitical landscape in the Middle East. Led by Cheney, Rumsfeld and Wolfowitz they fabricated lie after lie to deceive the American people and her allies to support a totally unnecessary and destructive invasion of Iraq in 2003. The consequence, as we all know now was a total calamity for the Iraqi people; at least 250,000 Iraqis were killed, many multiples of that maimed, and millions displaced. This war, which was opposed by so many people in the West, was also a disaster for the US and her most loyal ally Britain; over 4500 US and British soldiers were directly killed by the conflict, probably as many subsequently committed suicides, and at least ten times more were physically and mentally scarred for life. The arrogant and criminal actions of Western warmongers led by the neocon, and approved and happily indulged by the intellectually vacuous Bush and Blair created insufferable tragedies on all sides, and futilely created a far more dangerous and radicalised Islam abroad and at home[114].

But, the demonic act of bin Laden may well have prevented an even greater disaster for the whole world. Had 9/11 been prevented, as might have been the case, had the neocon not been transfixed on the idea of containing China, the US, through the demonic command of the neocon, would most likely have continued their original plan – of containing and if necessary instigating war with China. In this sense, China, US and the rest of the world (except Iraq) have been very fortunate to have escaped the hubris and criminal intentions of the US neocons. We can say this for sure because the irony of it all is, because the Bush administration was occupied with wars in Iraq and Afghanistan, its relationship with China in the end turned out to be quite amicable and cooperative. In fact, the noughties could be said to be the decade when China and America became an uneasy, but mutually needy couple.

### Chinese or English? The identity question

During the writing of this book, I had the idea of giving it a roguish title 'The invention of a Chinese Englandman' where the word 'Englandman' is a mischievous correspondence to 'Chinaman'. The latter is a pejorative word which I would never have entertained using. On the other hand, I am not comfortable with the more accurate description of being a 'Chinese Englishman' which is what my experience has made me. Ethnically, I am Chinese. But through my cultural upbringing, I am part Chinese and part English, yet I prefer my non-Chinese side being identified as British[115] rather than English.

So why do I feel so much more comfortable with the term British rather than English? The distinction for me is Britishness feels diverse and inclusive, whereas being English feels too specific, with perception that one has to be of Anglo Saxon origin, speak and act in a particular English way, or at least being born in England. There is an inherent inclusivity of being British, as it is a term used to describe the natives of Britain which is itself multi-cultural and multi-ethnic; the English, the Scots and the Welsh. Over their long histories, they have fought, mocked and even often antagonistic to each other, but somehow they have held together as a peaceful union for over 300 years, and subsequently even forged one of the most successful and stable nations in world history. The British Empire, while not something which I feel proud of, has nevertheless, also given Britain and Britishness an extra multicultural dimension. Of course, the 2014 Scottish independence referendum was a close shave for the unity of Britain. Personally, I was very relieved of the result which connects with my feeling of Britishness – that there is a sense of inclusive diversity, inherent multiculturalism, and tolerance in the status of being British – that we can all live harmoniously together without nationalism that drives so many nations and people apart. Therefore, I feel very comfortable with identifying myself as a Chinese Brit, or 'British Chinese' as is the more popular term of describing an ethnic Chinese with British nationality.

I am obviously a British Chinese by this definition, but what is the real nature of this identity? What does it really mean to be a Chinese Brit, or indeed of anyone cast with a dual identity label such as Black Briton, British Asian or British Jew? I shall try to explain my own particular situation based on my own feelings and experience; then I will further my thoughts to generalise to a broader spectrum of multicultural British society.

I'll start by defining the different states of a person's identity. Every individual is obviously unique with different character, belief and value. Much of these characteristics are influenced by the person's upbringing, environment and their genes. Thus, the root of a person's identity is really rather complex, dependent on many variables of life. Hence, it is necessary to be clear that the discussion of the identity issue here is restricted to the notion of the person's ethnicity, culture and nationality. For Black Britons, British Indians, British Muslims, British Chinese, British Poles, or any other ethnic British group, the identity consideration is how they fit into the majority British culture of indigenous White Britons[116]? In this regard, I have observed five classes of ethnic minority Britons: 1) those who are totally assimilated with the dominant majority culture of the country, with little interests and knowledge of their roots. They were predominantly born in Britain of 1st or later generation parents or with one parent of indigenous White British background, or raised by very westernised parents or guardians. Some members of this group may even reject or avoid association with their ethnic roots. By definition of their mind-set and upbringing, this group's identity is virtually like that of the majority British of the indigenous English, Northern Irish, Scot or Welsh. However, because of racism or bullying, it is possible some in this group will develop a dual identity over their lifetime. 2) A privileged group and some with considerable wealth who chose to make their home Britain. This highly mobile group is too small and elite to worth much discussion. The identity of this group is likely to be across the board, ranging from Anglophile to someone who only sees Britain as a convenient tax haven. 3) Older generation migrants who work in traditional and niche industries e.g. takeaways, restaurants, corner shops, taxi drivers. This group can be argued to have fitted well into British society as they provide an enriching service which the indigenous population needs or cannot provide. However, due to the long and isolating working hours in these industries, plus their relatively low educational background, these migrants tend to have little or poor command of the English language, have limited engagement with the majority culture except through their work or business. This group tends to be pragmatic and quietly gets along with life. Hence, despite their relative cultural isolation, the question of identity is a mute consideration. 4) Recent influx of predominately young adult migrants looking for better economic opportunity, or asylum seekers escaping wars or persecution. The migrants in this group are likely to be very grateful to be living in Britain and look forward to becoming British citizens. However, language and cultural barriers mean their identity, at least initially, or for a number of years is very likely to remain tied to their original roots. 5) A diverse group consisting mostly of second or higher generation British born ethnic minorities, and 1.5 generation[117] migrants with high levels of exposure to mainstream British education and culture, while maintaining a distinctive ethnic culture. Except when they were very young, English language is rarely an issue with this group as they get older. As British born or raised citizens, this group when compared with first generation migrants has a naturally strong self entitlement which makes them much more confrontational against racial injustice or prejudice. Difference in cultures and socio-economic circumstances can to varying degrees lead to barriers between groups of different ethnicity or religion. This group possibly possesses the most complex range of identity issues. While many have grown up adapting and appreciating their dual cultural heritage, some may feel ostracised and confused by the multiple influences of their surroundings. The vast majority of this group integrates well with general British society, although there are polarising issues with respect to socio-economic opportunities, and Islamist radicalisation of Muslim minorities.

As a 1.5g migrant, I belong very much to the complex fifth category above. This group is collectively the most diverse, their dual cultures and identities adding to the richness of

Britain in all walks of life ranging in arts, business, fashion, food, literature, music, politics, religions, sports, science and technology. The contribution of ethnic minorities has without doubt helped make Britain into one of the most creative and interesting countries in the world. The success stories of ethnic minorities, where they might not succeed elsewhere, are in my opinion part of the continuation story of Britain's tradition of openness[118] and her exceptional ability to foster creativity. It is with these two particular aspects of Britain which led to my feeling of association and pride. The development of my identity into a 'Chinese Brit' was a gradual process. I was granted full British citizenship at age 17, about 6 years after having arrived on this shore. I was happy to see my British passport, but the feeling of truly belonging probably wasn't embedded within me until my mid-twenties when I had made more British friends from the broader environment of University, the work place, and had fallen in love with classic British institutions such as the BBC, Radio 4, music, her great cities, seaside, genteel countryside, universities and research centres. But, why did it take this long to feel the belonging? First is the language barrier; learning to speak and interact in English fluently is not just a two year trial. No, not a five year trial either! I would say it takes about a decade of frequent engagement with fluent speakers to feel a sort of native ease and confidence with the language and culture. Furthermore, English people in general are reserved and impatient with people who could not speak the language well. To be accepted and integrated with the indigenous British culture, the immigrant has to try very hard and make the extra effort to join in. Secondly, despite the massive progress in race relations, racism[119] is still common which can sometimes really knock back one's confidence of being fully accepted, or wanting to integrate further. The final point is perhaps more innate, and a constant source of debate: is multiculturalism doomed to division between ethnic groups, and with the majority indigenous British population? My answer is an emphatic "No" because it is entirely natural to enjoy and indulge in one's heritage, and if one is lucky enough to be able to speak another language, people should feel free to use it. However, I do not think multiculturalism needs to be promoted or be funded by the government or tax payers' money. All people of Britain should share common ground, and reject some cultural traditions which are counter to the modern and liberal tradition of Britain. Indeed, I think the list for 'common ground' is just too long and obvious; most people of all races live a decent life with people generally helpful to each other in need. British society on the whole, as particularly demonstrated in its great institutions such as the NHS, is functionally harmonious between all the ethnic groups. However, on the other hand, it only takes a few lingering or persistent clashes of cultures to reflect badly between some ethnic minorities and mainstream British society. Despite massive societal improvement since my childhood, racism is still alienating many, especially in the age of social media which is sadly re-igniting common racial abuse previously damped by education and political correctness. Likewise, there are some predominately imported cultural practices that are obviously abhorrent; such as, female genital mutilation, baby sex selection for non-medical reasons, forced arrange marriages, marrying close relatives, disrespecting people of different beliefs, and most concerning of all, the spread of radical Islam. In the age of increasing anti-Semitism and Islamophobia, the insidious radicalisation of a significant minority of British Muslims leading them astray from mainstream values is a serious crisis, both here and abroad. A fundamental challenge here is, Islam, just like all other religions, is not perfect and needs reforming. Meanwhile, I think the onus is now on the liberal part of the Muslim community to actively fight against doctrinal Islamist ideology, while British law is strongly upheld to eliminate intolerant indoctrination and traditions. There is no easy solution, but there are general principles to foster a harmonious society for the whole country. It is all about education, education and education: to learn sound scientific teaching, sound humanitarian principles and sound ethics which is inclusive whilst not appearing arrogant. When we can all share strong common ground and obey just laws, then prejudice and injustice will be minimised. Personally, as mentioned earlier, I particularly admire and appreciate Britain's general culture of openness, inclusivity, diversity and outstanding creativity. Thus, by being able to identify oneself with the best of Britain, her character and achievements, rather than being embittered with the negative experiences that all ethnic minorities have endured; subconsciously and gradually, I became a Chinese Brit.

# Part II

# 8. AN INVENTOR'S JOURNEY

### Three Inventions, one dream

Between 2003 and 2007, I had a series of eureka moments which led to the development of three totally unrelated products. Invention number one was an appointment reminder device inspired by my regular hospital appointments. Invention two was a 'spectacle support system' to alleviate contact pressures from glasses, which ironically was partly inspired by the racist taunts ('flat nose', 'flat face') of my early years in Oldham. And invention number three was an intravenous injection aid, inspired by my own treatment struggle with needles. In this part of the book, I hope to convey the excitement and traumatic experiences which led to those ideas as well lessons of how to (or 'not to'!) make an idea to commercial proposition.

In the first chapter, I mentioned that it would have been very useful had the appointment card been capable of alerting me of an imminent appointment. Far fewer people would commit DNAs[120] and the NHS would save a significant sum in lost efficiency. Obviously, a paper appointment card will always be passive, and I didn't want to enlist expensive or bulky electronic devices to achieve the aim. Furthermore, I wanted a standalone solution which did not place additional demands on the patient nor require complicated user actions for it to function as an effective appointment reminder.

*(2003-2007)* 231

Subsequently, I looked at my wallet and saw the various credit cards that I used to juggle my finances. As is common even then, these cards, also generically known as smart cards, have 'chip and pin' features. They have microprocessor circuitry just like a mini computer which can be programmed via a small portable device very similar to card readers used in restaurants. Eureka!! Surely, my solution would be to create a 'super' smart card with the ability to tell the patient the time of the next appointment, the ability to sound an alarm to remind the card holder of an imminent appointment and be easily programmable via cheap and existing card-readers. But is such a "Smart Card Alarm" even possible? I wanted such a card to be as thin as a normal credit card (i.e. about 0.8 mm or less thick), because any thicker would have made it less pocketable, especially in male wallets, and it would not be compatible with existing card readers which are cheap and common. I felt if such a card were feasible, it would be a very effective product, with applications not just in healthcare, but maybe also across a whole spectrum of businesses which regularly give out appointments. I had to find out. The excitement of the eureka moment had turned into the quest for technical and commercial possibilities. Another motivating thought had also crept into my mind; if I could make this invention work, I would surely have erased my debt to the NHS!

God bless the internet – unlike my first tentative invention attempt in the mid-80's[121], it had become so much easier to find out if ones invention was novel or not, or to find out about other products, or where to find expertise[122]. So the first thing I did was to trawl through the internet to learn more about the technology of smart cards, the current state of the art, and whether any similar product already existed. Initial searches showed this appeared not to be the case. This was encouraging, although one has to be careful that it really doesn't exist because the search was not sufficiently exhaustive, the product may be technically too demanding or too expensive to get into the market, or it may already be under development by a major corporation or other enterprising start-up. Later on, I would often hear a common put down to inventors; there is no such product in the market because it is not going to sell, or there isn't a market for it! Unless these critics offered valid reasons, I refused to be fazed by their pessimistic comments usually dished out by people with limited understanding of the product and its market. If an inventor's new product satisfies these criteria,

    a)   It addresses unmet needs
    b)   The size of the unmet needs is reasonably large
    c)   The product is easy to use or implement
    d)   The cost benefit is reasonably significant relative to the cost of the product
    e)   There are new IPs in the creation

then, I believe it is a product worth further investigation. My imagined 'smart card alarm' surely satisfied the above criteria, but how could I make it a reality?

The reality was I had no experience of product development for any product, never mind a sophisticated compact electronic invention which no one seemed to have made before. Ironically, I was working in a University's electronic department at the time, but I was a theorist with zero practical electronics skill. Except being an expert patient within the NHS, I had no idea about how to get a new product adopted by the NHS. I think this is when most people with ideas gave up, whereas the real inventor continued to persist. I did not consider myself an inventor then because I only had an idea. Perhaps I could sell my idea, except of course an idea for a product is precisely just an idea. Companies can learn about your idea, dismiss it and then develop something very similar without even acknowledging you. However, a pure idea may be able to be turned into a valuable property via the creation of an intellectual property (IP) such as patents, design rights, copyrights and trademarks. Hence it made sense to convert my idea into a patent, a technical and legal document which sets out the rights of an invention. A few years earlier and the task of applying for a patent would have been a daunting prospect, conjuring up a complicated application process with a highly technical description and nightmarish legal jargon. And that is before one even considered the patent attorney's fees of about £200 per hour. However, the internet has demystified much of the fear. The UK Intellectual Property Office (IPO) provides very clear information about all aspects of patents and other IPs. It explains very clearly the structure of a patent and the cost of filing it, the latter costing less than £200 which covers the application fee, search report and examination fee. Since I could spare more time than money, I could even have a stab at writing my own patent. A patent basically consists of three parts; a full description of the invention, corresponding illustrations, and 'Claims'. Preparing the first two parts was relatively straightforward since I already had experience in technical writing and was quite good at drawing. It is usual to assume that the description must appeared to have been written by an expert lawyer, but a layperson that has followed the guidelines of writing such a document precisely, is also acceptable i.e. the writing style should be concise, relatively dry and terse – a style also trained to and expected of scientists. The trick is to describe as much as possible about the invention, adding any new features or uses which you can imagine in addition to ones of the initial main idea. Also, be as broad as is sensible about its construction and application. For example, I would not constrain the 'Smart card alarm' to be of credit card dimensions with an electronic paper display, embedded within laminated plastic, and to be used by doctor's surgeries only. This would have been foolishly restrictive as competitors could work round the patent with a slightly different sized device, or use a different display technology, or use a new type of non-plastic card. One then constructs the 'Claims' part of the patent document based on the description. This part gives the legal claims of the invention and is often versed in an extremely concise and legalistic manner, and hence is the one part of the patent application which I personally would not even try to draft. However, it is possible to file a patent without the claims section initially. As long as the claims section is submitted within the 12 months from the date of filing to the Intellectual Property Office, one can claim to have filed a patent claiming priority as the inventor on the date it was filed, which then allows the inventor to openly talk about the invention without fear of invalidating the invention[123]. After much research on the technology of smart card and related microelectronics components and circuitries; on 24th January 2004, I filed my very first patent application with the working title of "Activity reminder smart card". As I wrote the patent myself, omitting the claims section for the time, the cost of this application was just £30! I now had up to 12 months to get people interested in the idea because by the end of the year, I would need to find some serious money just to keep the patent going. The patent filing had brought me time and elevated my idea to a serious business proposition. I could now seek out advice or search out people or companies to discuss investments or a joint venture with greater confidence.

Looking back, 2003 was the genesis of not just the smart card alarm, but also another unrelated invention that involved a new kind of headwear. The concurrence of these two ideas led to a rather exciting but hectic Christmas and

New Year period as I prepared and then filed my second patent in early 2004. This patent has a rather technical title "Spectacle Support System" which basically describes new ways of utilising popular headwear (such as a sports cap) to alleviate contact pressure on the nose and temples of wearers of heavy or ill-fitting spectacles. This invention consists of a set of fully adjustable and folding support rods which can be attached independently beneath the peak of the cap, such that when they are folded down, a hook on the support rod can be engaged with the side of the temple arm of the spectacle, so that the weight of the glasses or spectacles can be transferred to the cap.

Why and how did I come up with these two inventions? Like most inventions from individual inventors, these ideas were conceived from first hand experiences. Most people typically don't give a second thought to the fairly minor but relatively annoying problems they encounter every day, or if they do think about them it is only fleetingly to wish someone else would do something about them. An inventor, on the other hand, tends to keep the problems they encounter in their head. Once in a while, they think of solutions to the problems. This may come through a mixture of systematic analysis and research, but more often than not though via a lucky moment of unrelated inspiration. However, just like almost any problem of the world, there is usually more than one solution. To narrow down the various possibilities, inventors need to consider exactly what they want their inventions to achieve.

The smart card alarm is an example of that process: I had personally missed an appointment and realised it must be a common occurrence faced by everybody at some time or another, and so I systematically looked for a suitable solution. The recognition that I could have used text message reminders didn't appeal to me because it made an assumption that mobile phone ownership or technological familiarity is ubiquitous, when in fact, it cannot be taken for granted by all sections of the community e.g. the elderly. In any case, the concept had already been developed with pilot trials already underway by others. I wanted a product solution which was universal, inexpensive, easy to program and distribute, and readily adaptable and easy to use by patients or carers. Having set out these demands, the possible solutions were significantly narrowed down, which after some research and maybe a serendipitous moment led me to the idea of a 'smart card' alarm.

For most of 2003, I had a full time job working as a research fellow in the "Physical thin layer group" in the Electronics department at the University of York. It was coincidental the group sounded like its research was related to my smart card alarm invention, whereas it actually refers to the study of the very thin microscopic surface layer of the material which determines its electronic properties. My job was to model the way electromagnetic waves (e.g. radio waves, light, radar pulse are all examples of electromagnetic waves) interacts with the material. The work has many applications relevant to telecommunications, stealth aircraft, and electronic shielding. The research work involved mainly developing computational modelling programs using a very popular and powerful technique of solving the Maxwell equations[124]. That technique is known as the Finite Difference Time Domain (FDTD) method, which by considering discrete electric and magnetic field components within a cubic cell structure turned the Maxwell equations into a discretised form capable of being solved by computer. The FDTD technique was invented by Kane Yee, a Chinese American in 1966. The method did not gain much recognition until the mid-80's when powerful computational power became more affordable making the method widely popular ever since. This is quite common in the story of innovations; many ideas or inventions in the past have often been ahead of their time, or just not picked up when they really deserved to. I had really enjoyed the work, because it was so motivating to understand such a fundamental theory and be able to apply the intuitive and beautiful FDTD technique. However, I did not enjoy the computational side. I had to learn a new programming language[125], learn all its 'grammar' and 'syntaxes'; artificial human constructs which didn't excite me in the same way as trying to learn or master new areas of fundamental physics or mathematics. Programming for me is ultimately rather tedious and boring despite it being so useful and powerful tool once mastered. There was also much uncertainty being a post-doc. My appointment was initially on a one year contract, and it was only a matter of a few weeks before the end of the contract that I learned the department had found the additional funding to extend it by one more year. This constant need to find new contract every few years is a common employment condition amongst science and engineering post docs in academia because permanent positions are few and so highly competitive.

So there I was, in a mentally demanding job which I really liked and disliked in almost equal measures, increasingly finding myself pondering ideas that had really started to grip the inventor and entrepreneurial sides of me. I haven't mentioned my H condition for a while, but at around this time, my knee and ankles were becoming increasingly troublesome, so with just six months left in my second year contract, I requested from the head of the research group Prof Marvin that I turn my current contract into a half time position. The request was accompanied, for the first time, by the divulgence of my medical condition. I had never previously divulged my medical condition directly to any of my previous managers due to my lifelong coyness of mentioning H. There were mix emotions of embarrassment and discomfort in the revelation. Internally, I was nervous as I needed to summon up courage for what seemed like a sinful confession. It was an emotional release though because my increasing bouts of limping at work didn't felt as strange afterward. My request for part time working was granted. As it happens, a part time position was ideal for both the employer and me: there would be more time to solve the work problems, more flexibility to recover from H related bleeds, and the time to work on my inventions whilst still receiving a regular income. The sacrifice of a smaller pay cheque was fine for me too as I was single, didn't have kids to look after, living in an economical one bedroom flat and quite frugal in all my needs anyway.

Working part-time was quite liberating; the pressure of work was still there as always, but there was more variety and excitement in the work as I juggled between academic research, looking for potential business partners and advocates for my inventions. Perhaps, without fully realising it, I had begun my retreat from the innocence of scientific academia to the hard world of business. Like many inventors and entrepreneurs across the land, the journey and dream of success had just got started.

2003 was an eventful year. After many years of apparently 'getting by', I started regular physiotherapy sessions in St James hospital in Leeds, to check on my knee problem, as well as, my increasingly troublesome ankles. I wished someone had told me to see them earlier as the brilliant (H-specialised trained) physios Ann and Thuvia could have significantly slowed down the insidious decline in my condition. Years of compromised walking on my knackered right knee had made me walk with an unhealthy gait for too long. Parts of my body were under undue stress – there were neck and back pains but most worryingly my ankles were starting to cause more grief[126] than my already knackered knee.

2003 was also an eventful year in world affairs. All of my personal trial was nothing compared to the nightmare and tragedy of Iraq. Most of the world and the great British public were against the drumbeat to war – Tony Benn gave an impassioned speech on campus, and had it not been for my disability I would have joined in the largest protest ever seen in London against the planned invasion of Iraq. The hubris and aggression of the American neocon and the appalling judgement of Tony Blair had led to incalculable suffering and instabilities in the Middle-East. As I previously speculated in the aftermath of 9/11, the conflict with Iraq could so easily been with China instead. In this respect, China and the world had been lucky. Meanwhile, China became the third nation in history to send a man to space.

## A year of searching and hope

The filing of my two patents in January and February of 2004 provided a good sense of satisfaction and relief. The months of intense churning of ideas and research had been successfully summarised into two neat patent documents, which could alter my destiny and possibly even made me rich! A patent is not just a techno-legal document, it is, in a sense, also a business document because one has to explain the invention's applications and uniqueness, justifying to the examiner and potential investors that you possess a truly novel and commercially tempting product concept. The process of writing the patent also helps to clarify one's understanding which could often leads to further technical and commercial ideas. Indeed, whilst writing the patent application I made several additions to the original core idea, giving the invention further claims which may turn out to be absolutely vital[127].

Without investment, there was no way that I could have made a working prototype of the Smart Card alarm. I had the idea and the concept, but to be frank I was not sure if it was really feasible. Can one really make a card with all the additional components (i.e. with the addition of battery, display, and speaker) which are as small and thin as a credit card? Would the battery hold enough capacity to be practical? Would the speaker be loud enough? Are there suitable displays and battery technologies to do the job? To have the patent filed was a first step and my only way to demonstrate my seriousness. It followed that I could then talk about my ideas more openly, and proactively seek opportunities to progress them further.

If the patent was granted it would also provide some insurance against competitors or being ripped off by unscrupulous people. Although I had been engaged in research for most of my adult life, I had no experience of how to take a product idea to the next level, or what that entailed. My last invention adventure in the mid-80's was a waste of money, except for some salutary lessons that ideas alone are not likely to attract the interests of manufacturers, and to be very careful about invention promotion companies. In 2004, I was a bit wiser and confident because English was now my dominant language and the internet information revolution had removed the many previously mysterious processes of taking a product to market. It allowed me to carry out independent technical and commercial research, which also enabled me to submit my own patent very cheaply. Also, I could use the web to seek out the people or companies who may be interested, be they potential investors, business partners or advocates and manufacturers.

There were also two other significant factors which made my re-entry to the invention and innovation world less daunting. The first was the significant number of often government backed support organisations which offer much free advice through business related events and seminars. They provide excellent theoretical and practical understanding about innovation and related matters such as access to finance, product design, prototyping, IP, and marketing. Some of them, such as Businesslink, Yorkshire Forward[128] and York Science City even offer financial support for innovation projects. These meetings also offer great networking opportunities, which I had not participated much during my physics research career or in my first entrepreneurial stint a few years earlier. As I later experienced first hand, much of business opportunity is about networking and about the chance encounter that might change your destiny. The second important factor which I am grateful to is the local inventors club. These clubs are formed by volunteers with varying degrees of expertise and experience. Many of these volunteers are inventors themselves with a helpful mixture of people with useful background in IP, manufacturing, or business. They usually meet monthly, invite guest speakers and enable old, aspiring and new inventors to mingle and offer each other advice and support. I feel particularly grateful and privileged to be a member of Ideas North West which is based in Blackburn, and the Manchester Inventors Group. These clubs have provided valuable advice and inspiration ever since I first joined them a decade earlier.

As an aside diversion, it is interesting to reflect that clubs and societies run by volunteers can really make a big positive difference to the well-being of society. These clubs are run by enthusiasts who love what they do. Although volunteering is not a uniquely British trait, I noticed that Anglo-Saxon society has a particular strong and distinguished record of inventing famous benevolent clubs such as the WI, the rotary club, the boy scouts movements, the institute of advanced motorists etc. As mentioned previously, another volunteering club which I have a great fondness of is the York Advanced Motorcyclists club. Besides learning life-saving skills, I have had enormous fun and camaraderie during the 10 years which I was a member. These clubs are a reflection of an advanced civic society where people frequently give their time and share their expertise for free. Every time I participate in club events, my affection for the club, and its people is increased, which in turn filters into my consciousness a wonderful sense of being part of a bigger community. Personally, I found patriotic education to be nonsense, whereas active participation in benevolent and non-exclusive communities is what binds people together and identify my own personal admiration and attachment to being part of the British society.

The internet research and various networking events provided me with some hopeful contacts from several key NHS people, which included the local York hospital chairman Professor Maynard who was also a health economist at the University. Another advocate was a medical device innovator from a nearby hospital. They offered me approving comments which were very encouraging, but it would be naïve to think these would lead directly to concrete financial support. However, their support could lead to more open doors and act as an advocate or champion of the Smart card alarm. My other step was to find financial support and investment, which yielded some hopeful appointments, one from an East London NHS Innovations hub, and two business angels. These small steps are just the very beginning process of innovation, but at the time they were exciting prospects and endeavours as I knocked on unfamiliar doors, met new people of distinctly different professional backgrounds and encountered new or as yet unknown places.

All investors have one common objective; to make a profit. The higher the risk, the higher their expectation of the profit returns. They don't necessarily care about the technology; it's all about the idea which they have to like, and whether you have sufficient competitive advantage in the business. In my case, a good IP (i.e. the patent) position combined with the strengths of the business case (i.e. whether there is a viable market for the product) would mitigate the risk. One business angel, who made his fortune in the spice trade dismissed the smart card alarm quite early on, as he thought the future would be all about the mobile phone. I appreciated his point, but disagreed politely. I offered my explanation as usual…yes mobile phones can do many things, but that doesn't mean there is no room for other products. It is disappointing when potential investors don't agree with you, but reasonable objections to ANY idea will always be forthcoming, and knowing them in advance enables better preparation for further encounters. Of course, one can also get ignorant or unreasonable objections to one's idea. This shouldn't be taken to heart since a business angel is just another human being who can be just as stupid or wrong as anyone else.

The two other potential investors liked the idea, so they turned their attention to my IP, namely the patent. Within matter of days, one of them discovered a product called 'ePass' which had very similar features to mine, except it was much bulkier and could not be programmed by a standard smart card reader. Hence, my concept appeared not to be new and therefore some part of my patent must be invalid due to the existence of 'prior art'. The other investor also did the check, this time with the help of professionals who were experts in finding patent prior art from the huge patent databases. The results of their search revealed several patents which when combined together apparently had all the main features of my patent. As I checked through these patents carefully, I felt increasingly disheartened when key phrases and words such as smart card, display, compact power source, programmable etc. started to pop out of their pages. On the basis that my IP may not be as unique or as assured as I thought, both potential investors checked out and wished me good luck.

The programmable compact alarm is therefore definitely not a new idea, but the reality was no one had made a truly credit card sized Smart card alarm with all the features previously mentioned. I knew my patent definitely had other new features and enabling technology which were not mentioned in any of these prior arts. These features were embedded in the detail, but most investors would not have the patience or the expertise to comb through such technical documents. It was quite painful to confront these prior arts even though not all of them were surprising since I had carried out searches prior to my own patent write up. I was forced to think up new useful features or enabling methods and technologies which would ensure that if the Smart card-alarm were ever made, it would be more effective and useful than any existing known description. These features included an innovative use of optical fibre for a visual alert when the card is hidden in a wallet, and a specific technique to ensure that the limited battery capacity could provide a sufficiently effective audio alarm.

So there was some positive news after all. I got to understand the market and product situation much better; no one else had yet made a real smart card alarm, and there was probably still some potentially useful IP to derive from my patent. However, the negative of the weakened IP position raised significantly the uncertainty of being able to get the necessary investment to move forward. Also, why had no one else made such a card yet? Was it feasible? I don't recall if I have even contemplated giving up at this stage. As an optimist, all of these obstacles didn't deter me. It felt natural just to keep ploughing on. I would not abandon my patent application or the attempt to create a truly innovative product which could really make a significant impact on the wasteful 'DNAs' (no-shows) problems in the NHS.

Whilst I was chasing opportunities for the smart card alarm, I was simultaneously pursuing the same goal for the spectacle support system idea – a specially designed cap (known as the "iiCap") which eliminates contact pressures due to wearing heavy or ill-fitting spectacles. Again my patent was my key asset; although this time I had also managed to concoct several prototypes which were functional but Heath Robinson-esq. This invention was also revealed to the above investors as well as several companies. The outcome was similar. Some rejected the opportunity because they didn't get it, and others thought the idea was too easy to copy, and unpatentable because of apparently similar prior arts, or even as one famous designer thought, that it was a great idea, but it may be too simple to be patentable! As before, they missed important details or misinterpreted the prior art. From a market point of view, encouragingly, it impressed some key people including an eminent surgeon I met at a medical innovation networking event in the Manchester City stadium. However, there were also equally unimpressed people who thought it was just a pointless gadget. These people would say it's not relevant because glasses are very light nowadays, people could wear contact lenses instead, people won't wear a cap or they could have laser surgery. These alternatives are all facts of course, but why focus on the non-buyers? Somehow they miss the point that billions of people wear glasses or caps or both and so even if only a small proportion of these people are interested, there would be a substantial number of potential buyers. It seemed like investors tended to focus on the negatives, basing their investment decisions primarily on propositions which have least negatives, rather than looking at the positives of the opportunities, and weighing up the balance between them.

These rejections were annoying but didn't dissuade me much because rejections are not unusual and it was to be expected. None of the objections encountered so far were conclusive, although the patent position was worrying. It led me to think that investors sometimes rely too much on the patent position rather than the market opportunity and further potential to create other new IPs other than patents (e.g. design, copyright, trademarks and exclusive propriety technologies which can be kept secret). A patent is very useful and potentially very valuable if you can get it, but it's not a prerequisite for success or even always necessary. There are different ways to make money from a new idea. 1) tell your idea to a company, with a confidential agreement in place and hope a company will want to license it, 2) obtain some official IP rights via patents, copyrights, design registrations, trademarks etc., and then just sell or license the IPs to suitable companies, 3) create the product and get it to market yourself, with or without IP rights, 4) copy other people's ideas. The first situation is extremely rare since most companies will not consider idea submissions from people without them first having secured some official IP. However, having the good idea could be the springboard to attract potential partners or research grants which could advance the idea further. The second situation is what many inventors wish for – being the owner of some IPs licensed to a successful manufacturer or brand that would professionally design and produce your products, and then market them. The inventor would then sit back and just enjoy collecting the quarterly royalty cheque!

Indeed, I was hoping this would be the case for me. Getting IPs such as trademarks and design registrations are actually quite easy and not that expensive, but these kind of IPs are usually only valuable as licensed objects if they are already successful or as a combination with a patent. Getting a patent granted can take up to 3 or more years and can become prohibitively expensive for the individual inventor. Furthermore, you will most likely succeed when there is a prototype, which again can be very expensive. As for the wishful royalty cheques; unless the products sell in very high quantity with large profit margins, you are unlikely to make much money either. Hence, one would find that most often, many inventors end up following the third way of trying to create the product and take it to market themselves. In fact, it is usually their only way. It is the most risky route, but also the most rewarding if successful. Investment is almost always needed. Come to think of it, virtually all the most famous companies in the world were created by people who founded their company in order to realise their visions for their products or services[129]. The fourth way of making money from a new idea is to copy it. This is one of the main reasons that some investors emphasise so much about IP and patent in particular, although the same reasons are also used to refuse investment on the grounds that people can copy the idea and, therefore, the patent even when granted is useless[130]. Generally though, copying and adapting other people's idea is probably the most common type of technology start-ups. Here, I don't mean the competitor simply copies the originator's idea, but rather they followed the concept or market leads initiated by the original innovator using possibly licensed technology or technology they built themselves. The originators have first mover advantage but the followers may become more successful than the originators as they can avoid the mistakes and cost of opening up a new market. Thus, they might invest more wisely and subsequently have more creative energy in the implementation than the originator. Classic examples are Facebook vs. MySpace and iPhones vs. Nokia. In a small and premature way, I feared most investors shared similar anxieties as both the Smart card alarm and the iiCap were original ideas which needed to open up a brand new market.

Over a period of several months, I had transited from situation 1) to situation 2), and was becoming increasingly drawn to situation 3) which was the most likely route to turn ideas into real products. I didn't wish to go through this route though, but circumstance dictated it, as in order to realistically get investors interested I would need to find out if the technology was feasible or not. How much would it cost to develop and would the market price make economic sense? Searching through the internet, there were many companies which manufactured ordinary smart cards, but only a handful of companies which seemed to be working on integrating compact electronic components within a smart card form factor. After further research and attempted communication, only one company responded positively – a Paris based company called Audiosmartcard. The name of the company sounded worryingly like my idea. However, closer inspection of their products indicated they were in the area of making high security smart cards which emit an encrypted sound over the telephone line as a form of additional authentication. Their latest cards had the same thickness dimension (approximately 0.8mm) as a normal credit card, and yet incorporated its own power source, a speaker and a beefed up microprocessor for handling audio authentication technology. My contact with a foreign technology company at the time was a major departure from my insular existence. Due to my mobility problems, I had not been inclined to travel much with my last trip abroad almost a decade earlier. I felt excited that a high tech French company would express interest and enthusiasm for my idea. I also felt nervous that they could lose interest soon when they realised that I was just a one man inventor. My confidence of dealing with this company was due to the excellent rapport with Philippe Blot, the VP of sales of the company who courteously answered all my queries, and showed exceptional openness to my concerns and probing. Of course, the company was particularly interested in my idea because they wanted to sell their expertise, but Philippe was not just a charming and superb salesman, he acted more like a business mentor who provided me with really useful advice and technology insider knowledge. I was convinced that further technical and business development would result with his help.

It was still necessary to raise funding to enable the project, although I felt my chances had improved significantly with Philippe championing my idea to his industry's contacts. Indeed, within a few months of our first contact, he got an approving interest from one of Audiosmartcard's clients in the USA. That company was InCard technologies Inc. (or 'InCardtech' for short) whose CEO Alan Finkelstein expressed to Philippe a desire to help me develop the Smart card alarm. This was wonderfully uplifting news. It gave me a real boost of confidence that obstacles which previously seemed so insurmountable appeared to be ebbing away. It was now up to me to chase this lead and clarify what kind of assistance I could get. My contacts with this Los Angeles based company were Nick Leung and Thep Tan who were responsible for the project management and business development side of the business. From their names, they were obviously Asian Americans of Chinese extractions. Although race and ethnicity is generally irrelevant to doing business in today's multicultural age, the fact remained that when in an unfamiliar environment, there is inevitably a small tinge of comfort from dealing with people who share a common identity. Perhaps they would help champion my idea to their boss, whom incidentally also matched my often absurd ability to connect the nonpertinence; Finkelstein is Jewish, kind of sounds like Einstein my hero, so may be my destiny was to meet him! It was important that I kept them interested by frequently updating them of my progress on my research into the business model, contacts I had made with the NHS and the health industry, and my attempts to raise finance. This enabled them to see that I was serious, and that their time helping me could be worthwhile to them in the long term. On the other hand, I needed clarification to what extent Finkelstein meant by his offer to help me. Emails started to flow between Nick and later almost exclusively Thep and I over my product idea, business plan and clarifications of CEO Alan Finkelstein's intention. Frustratingly, the CEO never answered directly and replies from Thep were brief and often delayed due to them being busy with other commitments. I felt it was not possible to push them too hard whilst they were genuinely busy. It was a fine balancing act of regularly nudging them until I got their confirmation, whilst not making a nuisance of myself. Hence, it was with great relief that almost 3 months after first contact, I received a reply from Thep that his boss Alan had confirmed his willingness to manufacture me some sample cards at no cost subject to an agreement being in place! I remembered this email well as I read it at late evening on that memorable date, 30th August. Despite the uncertainties in the statement, I felt quietly jubilant that the project had gained some further substance. However, this was just the beginning of a new set of enquiry and clarification about the agreement and what he meant by 'sample cards'. How would they manufacture them? Was I supposed to give them the specification which I wanted to achieve and would they then just manufacture it? There were still a lot of queries as I had never been involved in any industrial design of any products before, never mind an innovative piece of hardware technology requiring bespoke embedded software which must also be compatible with existing card technology standards. Although Thep was a great guy, I sensed he was under a lot of pressure from work and he could only relay my request to his boss at the right moment. Frustratingly, the situation was still as fuzzy as ever as I could not get hold of the CEO by phone as he was always at meetings or busy travelling. Frankly, I was probably too timid to really nail him down in case he changed his mind due to my irritating demand. It would be better to see him face to face.

Meanwhile, I remained in contact with Philippe to further my understanding of the industry, and may be through him to find out more about this so far elusive Finkelstein fellow. It was time to go beyond emails and conference calls with his subordinates. For the first time in almost a decade, I was compelled to travel abroad, a no big deal action of modern man except I carried more anxiety and burden due to the unpredictable nature of my knee to sustain the whole journey. I would travel to Paris in November to attend the smart card industry's premier tradeshow to get to know Philippe better, while more importantly to pin down Mr 'Busystein' for a definitive commitment schedule.

## Hello Dragons!

Amidst all this chasing and relationship building with my potential foreign saviours, I was also hunting for funding too, knowing full well that income from my University job would cease at the end of September. I was only mildly anxious as I had built up a reasonable amount of savings of around £12K, and my small flat had gone up in value significantly. Furthermore, the long and bitter battle for compensation from the H-community with respect to the Hepatitis C virus infection was finally resolved[131]. The government had finally set up the Skipton Fund to administer ex-gratia payment to all H who were infected by HCV. Unfortunately, or fortunately, depending how you see it, I was entitled to £20K compensation which arrived a few days before the end of September when my salaried job ended and my new career as a full time inventor entrepreneur began! This sum offered me a small comfort zone, but I knew I needed a lot more to create a technology business which required both building hardware and software. A month earlier, a friend emailed me a link about a brand new BBC program where contestants can pitch their business ideas to real investors on TV. Although I did not have much information about the program, it was worth a try as it was free to enter and a great promotion opportunity even if I didn't get the investment.

At the time, I could have submitted either my card or iiCap invention as a proposition for the pitch. Both were at a similar stage of development or 'under development' as might be a better description then. In my simultaneous search to develop the iiCap further, I discovered an award winning hat designer who initially really liked the idea and wanted to help me just like the champions that I had found in my card invention. However, she did not follow up with her promise to help out when I suggested entering her for the Dragons' Den competition. For some reason, that mere suggestion seemed to have filled her with such dread that she refused further contact with me again! I found it really unsettling that apparently friendly people could suddenly go completely silent, ignoring emails or phone calls entirely with no explanation. I did not realise at the time, but I guess this was the business partnership equivalent of being dumped. Why can't people just express themselves frankly?

So, I filled in the BBC application form using the card invention as my proposition. Although the path to developing the card had built up some momentum, I had essentially just an idea and a business case to write about. Within a few days of my submission, I got a call from the BBC asking me further probing questions which I readily answered as I had been fine-tuning my sales pitch since the beginning of the year when I started looking for support. Within a week, I was invited to the BBC Manchester studio for an audition. Having never done an audition before I had no clear idea of what an audition was, or the role of the various people who greeted me. They asked me further questions and then gave me instructions to present a pitch in front of a camera. Although normally quite reserved, I can be quite ebullient and combative when talking about things which I am interested in. This was no time to hold back, and I did not. I also tried to demonstrate the principle of how the smart card alarm could work by pre-setting my mobile phone to emit an alarm in the middle of the interview. Everyone laughed and got the idea when I explained that the alarm alert was from my pretend smart card! My dramatic presentation must surely have guaranteed me a place on the show. Well, the senior guy Graham Hoyland assured me that I had done well but couldn't give a definitive answer just yet. I went home feeling satisfied wishing there was a good chance that I would be presenting my pitch to some investors soon.

I did not have to wait long for the call. But it wasn't the news I was waiting for. The call was from one of the show's many researchers. They wanted to see and know more about my patent application. That's fine as I was open about it, although I feared that any initial search for prior art may give them a false negative expectation. Then, a few days later, they requested to see the business plan. At the time, I didn't have one, so I developed a detailed one from scratch, even though there were still so many uncertainties. Writing a business plan was a good idea anyway, and gave me further hope that the programme people must be increasingly interested in me. However, over the following two weeks, there were almost daily calls from different researchers asking me more and more questions. Some were about the business propositions which I had expected, but there were also many personal questions asking about my past, my hobbies, previous employment etc. etc.! At first, it made me feel quite good that someone was interested about my life, and this must be a precursor final check before the final "YES" answer. Then there were more calls; this time about my financial past, if I had been bankrupt before, or had a criminal record. The questions even extended to asking about my family, asking whether any of them had been on TV before etc. BLOODY hell… I wondered: when are they going to stop this flipping inquisition and give me the bloody good news? Surely I must have gotten it after all this! By the end of the third week of this inquisition, Graham, who turned out to be just an associate producer of the show finally gave me a definite answer: "we will know when I show your audition tape to the producer in London later this week".

I couldn't believe it! After all the preparation and interrogation over the past month, the producer had apparently not even been involved yet! All of my anticipation and excitement could be extinguished suddenly and what a disheartening feeling that would be! I didn't know at the time, apparently there were over 5000 applications, and only about 70 people would reach the final televised stage of the show. The people who interviewed me on the phone were just junior researchers, part of a team making up the show consisting of researchers, a director, an associate producer and the producer. At the time, I had little understanding of programme making and the roles of these media people. The fact that someone from the mighty BBC repeatedly wanted to find out things about me had made me feel optimistic and believe it inevitable that I must have been very close to being selected. With so many applications, the chances of getting on the show would not be down to just the merit of the business case; factors such as likeability or being telegenic were probably equally important. The reality was I was just one of many shortlisted by the BBC production team. There was still one last hurdle of overcoming the whimsical mood of the main producer down at the London headquarters.

That call, THE CALL from Graham finally came. "Kin, you made it to the show" or something to that effect were his words. "That's absolutely brilliant" I exclaimed. I thanked Graham profusely as he was very supportive, and no doubt had championed my case to the producer. After a month of building up the expectation, it would have been very deeply disappointing had I not been selected. This wonderful chance to advertise my project to the nation and a possibility to get investment from obviously very successful business people was a once in a lifetime opportunity. Some people with the benefit of hindsight about the program may think otherwise, but the filming was for the very first series of the show where nobody really knew what to expect. My filming slot was in two weeks' time on the 19th October. What a wonderful and exciting way to start my first week of being a full time inventor cum entrepreneur! But, that fortnight prior to the filming turned into my most intensely angry and resentful period of my life. When I received the good news from Graham, I did not immediately tell anyone about it. I just absorbed it, happy and relieved that all that effort and expectation was not in vain. Later that day, I told my mum of the news "I made it into this BBC program… I am going to be on TV!" She was busy doing some chores at the time and just kind of nodded her head and muttered some brief words of acknowledgement. I was disappointed and annoyed by her indifference but didn't think too much about it as she may have been distracted at the time. I would wait until dinner time to divulge my big news to both mum and dad later that day. I tried to pick the best moment to let them know, not distracted by chores or the TV. I repeated the exciting news again, this time with a bit more celebratory tone to reflect how exciting and important that news was! It's nice to share good news, but instead of responding with at least a mild approval, my usually reticent dad in an unexpectedly sharp and angry voice criticised me forcefully, apparently disapproving of my show of merriness, and devaluing its significance. My mum likewise followed my dad with a call to cease what to them was my boasting over nothing. I walked off fuming and bedazzled that we couldn't celebrate some good news together. I did not expect them to jump up and down about it, especially my dad who doesn't easily show his emotion. However, to get shouted down completely depressed me about ever wanting to share any bits of good news with them. It would not be up to their standard! Could they not just say "congratulations" or "well done" even if it was just a little success? Is it only ok to show or express congratulations on big successes only? What success is good enough to deserve their attention? Why are Chinese parents so emotionally constipated? My anger and disappointment with their reaction probably caused the bleed in my already vulnerable knee which further exacerbated my negative mood. My supposed day of joy and deserved mini celebration had turned completely sour. I felt anger and bitterness for the rest of the week. And it lingered on. They had deprived me days of high and exciting anticipation that one would have felt prior to the filming. For the first time in my life, my feeling grew into intense resentment of them and their attitude.

My mood was evident to my parents, and they expressed regret for the hurt. I tried to forgive them and returned to normality, at least on the surface. On reflection today, I can see how some children and their parents could suddenly break contact. It could be over a relatively small argument or wrongful criticism, like my experience above. However, the consequence of some argument, particularly at a sensitive time could sometimes amplify into a prolonged period of anguish because apparently trivial disagreements may well have ruined a good moment or event that may never be repeated. That loss can in turn lead to deep resentment. Fortunately I am not a begrudging person; I know my parents love me unconditionally. I tried to understand their culture, their generation mind-set, and tried not to compare their old fashioned and sometimes stoic attitude with the idealised image of an enlightened (and frequently denominated by western-values) family where everyone is attuned to each others feelings and where emotional expression is free flowing. However, that unusually deep resentful hurt feeling I felt did not dissipate quickly. For a long time, I felt I never wanted to express any news of interest to them, to punish them as though they don't need to hear any news from me since they won't appreciate it anyway. Even on my return from the Dragons' Den, I held back from talking about my experience with my parents.

I would need to prepare hard for the fateful filming date. The pitch must be less than 3 minutes long, which I thought was ridiculously restrictive, although with hindsight it made sense for television, and each minute in front of the Dragons seemed at least thrice as long. My prepared pitch started with:

*"Hi Dragons, my name is Kin Fai Kam, I am an inventor and a physicist…."*

It was a great opportunity to announce to the great British public who I was and introduce the product which could save the NHS millions of pounds annually. I was quite deliberate about the above introduction, wanting to let the audience know that I was a proper scientist as well as an 'eccentric' inventor. One thing which I discounted immediately was putting my PhD 'Doctor' title in front of my name. That would have been like waving a red rag in front of angry bulls. The use of one's academic title in that environment was pointless. It would have created an educational status-barrier which would have been likely to inflame the less academic Dragons – with retaliation against the contestant for their ivory tower ignorance of the 'real' world. I recall vividly the fate of one such contestant who dared to introduce himself as Dr so and so, who then got one of the most painful and humiliating put-downs of the entire series. "You are a Doctor? I actually think you need one" exclaimed cruelly Peter Jones to this poor guy who responded with his head downtrodden, whilst biting his lip. Mind you, his product was rather ridiculous; a PC with a wooden casing which looked as though it had been knocked together by a 10 year old apprentice carpenter.

The first series of Dragons' Den was really filmed in a dungeon like derelict building in Stamford Hill, North London. There were six contestants in the day of filming. We left our hotel early in the morning, ferried in two separate people carriers, driven past some famous London landmarks, and eventually to this unremarkable suburban street with shops and terraced housing. The passengers and I were all chatting excitingly about what was to come. When we arrived we were met with familiar and friendly faces of the researchers who had previously so deeply probed into our private lives. They kindly served us hot beverages and a full basket range of pastries, which was rather delicious to my greedy mouth. But later on, we were served with a pile of documents where all the contestants had to sign in order to continue. There were some confidential non-disclosure agreement forms and a release indemnity form. Also there were complicated forms which waived all claims of copyright to the material being filmed and other legalese about the rights of the BBC to exploit it however they wish. These were not simple light documents. I thought it was unfair that we contestants were subjected to signing such non-trivial documents in this manner on the day, instead of prior to coming to the show. I don't think I or any of the contestants seriously had the opportunity to read through them properly before we signed away our rights. I felt somewhat exploited at this point.

I can recall all the contestants well. One was a businesswoman from Cheshire who appeared to be already very successfully manufacturing and selling toys for her company. She wanted investment to develop and sell a new toy which her firm had invented. Speaking to her in the same platform with other serious business people somehow made me feel that my destiny had truly changed. My days as a desk-bound, reading, thinking and programming scientist were truly over! However, I didn't quite understand why she needed such a relatively small investment when she was already established. Another contestant was an inventor with an electronic locking mechanism which prevented doors from being pushed over by robbers or unscrupulous intruders. He had already developed an actual working product (not sure if it was just a prototype, but it was certainly very deep in the product development cycle), had granted patents, trademarks and several letters of endorsement from a police chief and several council authorities. His accomplishments were very impressive compared to what I had at the time, which was basically an idea, a patent application and just some relevant contacts who might (or might not) be able to develop my idea. However, I noticed one thing which perturbed me; some of his documents were over 6 years old. He must have been trying for at least that long to get to where he was now. Six years sounded like a hellishly long time. The previous 12 months were arduous enough; I could not imagine persisting for so long and hoped my invention journey would end in success much sooner. He was in the den for a very long time, may be as long as one hour plus, as I waited impatiently for my turn in the cold and draughty green room. He must have raised the interests of the Dragons. There was only me and one other contestant left by mid-afternoon. I was to be the penultimate contestant of the day to enter the den. The last contestant was a young lady from London with some fancy jewellery designs. Of all the contestants, she was the least amicable, the most intense, and kept herself to herself. Perhaps she was just shy, or maybe she was just being very shrewd and focused. Anyhow, I was here to do a job, not to try to chat her up!

Finally, I was led upstairs, not immediately to meet the Dragons, but to the makeup room. I know this is a common procedure in television, but it felt totally strange to have makeup put on my face for the first time in my life. I met Graham again, he told me that no one had got an investment today, and the Dragons were in a good mood. He was very encouraging, and I felt his sincere warmth and best wishes for me. Perhaps this is going to be my day! I was itching to go; I felt ready and was clutching my business plan ready to answer any tough questions from the Dragons. But one of the production team immediately forbids me from taking the business plan into the den. This was unfair I thought, but it wasn't the time to argue. The producer also finally revealed himself at last; a tall cool looking guy with straight long blonde hair who looked startlingly like the Hollywood actor Julian Sand who often played posh English characters. I thanked him for giving me this opportunity. He wished me luck; just at that moment I noticed a terribly steep staircase which I had to surmount to meet the Dragons. I immediately asked if I could use the lift instead as I had not expected this! Their response was a hurried no as it was part of the filming, part of an entertaining plot which I hadn't realised then. I suppose they did not know that I had such a severely damaged knee as I had not told them earlier. I immediately calculated in my head whether my bad knee could get up those stairs without looking crippled and discombobulated. To my surprise, I made it upstairs, just! But going downstairs would be considerably more stressful and I would have to hobble down using my left leg only, one step at a time. It would be embarrassing and humiliating for someone who didn't wish to be seen as a disabled person, even if I am one.

So, even before I finally got to glimpse the unknown beyond the stairs, I had to battle a major anxiety moment. Since this was the very first series, all the now familiar scenery which appeared in front of my eyes was completely brand new and unexpected. I noticed a flurry of activity, of cameramen, technicians, and presumably the director in the background, which rapidly faded away as my attention turned to the five serious looking people directly ahead of me, staring at me as I approached the spot which I was supposed to speak from. My immediate impression was the oppressive looks in all of the Dragons, except the only female Dragon on the panel. She had a soft and gentle face despite trying to look cross, which somehow soothed me just a bit. The Dragons were

Rachel Elnaugh, Duncan Bannatyne, Simon Woodroffe, Doug Richard and Peter Jones. At the time, none of them were famous but because they were all supposed to be highly successful business people, sat grandly in their business suits with cash strewn on their sides, they exuded authority and power. Their decisions over the next short period of my life could make a profound impact on my future. But to be frank, I wasn't that intimidated by them as I have the self-confidence of someone who has experienced too many real fears before. I had wanted to demonstrate my calmness and professionalism in delivering my pitch, but instead I revealed one of my hidden weaknesses literally from the word go! As mentioned in the beginning of the book, my first attempt at my pitch was a total farce as I unexpectedly gasped for breath after the stress of hobbling up and down those stairs twice on my precarious limbs. In fact, my physical situation could have been worse still, the night before in the hotel, my left elbow, which was another problem joint, had suffered a recurring bleed. Fortunately, I had brought with me some Factor 8 treatment which I administered that night thus prevented it from getting worse. Had I forgotten to bring the treatment, the Dragons would have witnessed the contorted face of a contestant experiencing the worst pain intensity and stress not even they thought was possible to inflict.

Eventually, the second time round, I managed to finish my pitch smoothly. My initial stuttering start was all forgotten as I prepared myself for the questions. Rachel Elnaugh asked the first question. I think it was about profit margin for the business. I couldn't answer immediately because I didn't have all the costings yet. I couldn't give a realistic answer for such an early stage technology project, although it would have helped if I could have accessed my business plan. I could have given an estimate based on various assumptions. It was not the best of starts. Next came Doug Richards. He was the most serious looking guy with a serious American accent. For some reason, with all their wonderful movies and jovial nature that Americans are famous for, when it comes to business Americans always sound to me the most authoritative and impressive. Perhaps it's their world dominance in so many fields which rubs off into their business brethren. He sounded the most reasonable though, politely asking for a demonstration of the card which I was holding during the pitch. Of course, the Smart card alarm had not been made yet. To demonstrate the concept, I had borrowed one of the authentication sound cards from Philippe Blot to demonstrate the principle. Unfortunately, the power in the card's battery had run out, and no sound could be heard. The heat on me was getting hotter, metaphorically and literally as the bright studio lights shone warmly above my head. I could feel my brow was sweating at this stage. Then came Duncan Bannatyne, who looked the most arrogant and cocky of the five Dragons. He carried a contemptuous expression, backed up by a disdainful tone made particularly harsh by his Glaswegian accent. I cannot honestly remember his comments or questions, except for his delivery which wasn't kind. There was a moment of silent pause between each Dragon's judgement. Then, with a flurry of seemingly irrelevant objections, Rachel Elnaugh declared herself "I am out! I am not going to invest in you" or something to that effect – I cannot recall the exact quote. It was the first time I heard the now infamous rejection phrase. Rapidly, further rejections barbed with even more bitter venom came quickly from Doug and Duncan. I have never experienced rejections like this before; in such rapid succession and entirely focused on the negatives. I noticed then my eyes went a bit misty from the potpourri of emotions that I was experiencing at the time. It's not like I wanted to cry or anything close, but I became astutely aware that my eyes felt moist, and my vivid imagination was playing through that incredibly embarrassing possibility that tears might soon be dripping from my eyes. An inner thought came to me at this point "What the f**k am I doing here?" and please "don't cry". Please control my tear ducts.

Fortunately, the intensity of the event didn't allow me to imagine myself to such a pathetic situation. Peter Jones now raided in with his judgement, "Everyone now has a mobile phone… your product is irrelevant". I responded with an uncharacteristic bluntness, calling him arrogant. He was visibly annoyed and then proceeded to lecture me about the number of mobile phones in the UK. I was glad that I ruffled him, but by that time, I felt my chances of getting the investment were rapidly diminishing. Like a live play, the outcome of the ending was getting clearer by the second. I did not bother to fight my corner. Despite his angry retort, there was something particularly impressive and likeable about him. I think it was his youth which impressed me subconsciously; that someone who was not much older than me could be living in a completely parallel universe of wealth and success. Funnily enough, I had not noticed his gigantic stature at all. The final dragon to pass judgement was Simon Woodruff, the cofounder of Yo!Sushi. In my view, he had the least gravitas of all the male Dragons. There was something flippant about him, in his delivery, and in the things he said. In fact, I cannot recall anything of business substance he said, except the final blow of "I am out!"

Dazed, I hobbled down the staircase for the final time, feeling even more embarrassed as I saw the faces of the people who had wished me luck not long ago. I felt I had let them down even though I felt an overwhelming sense of having just been exploited. It was truly a whirlwind experience. Disappointed, dejected, dazed, feelings of exploitation, this was not what I had bargained for. Given the title of the program, I had expected a confrontation based on business judgements rather than being made to feel exploited for entertainment. I put on a brave face for the final part of the day's filming, an interview by Evan Davies, the then economic editor of BBC2's Newsnight which happened to be my favourite news programme. Newsnight seemed to have a knack of employing the best economic journalists and editors, and Evan was indeed one of those great journalists whom I respected a lot. His face was familiar to me, but it was rather amusing to meet him face to face. I was startled by his very small head which was balding and angled sharply upward like a bullet. With his lanky figure, and his large eyes being so closed to each other, Evan Davis was the closest human being that I have ever met which resembles an alien. Fortunately, his demeanour was kind and gentle and I felt a lot freer to express myself, even cracking a few laughs amongst the crew and taking the opportunity to advertise my other invention (the iiCap). As soon as the interview was over, I was hurriedly chucked out of the building, into a waiting taxi back to the train station.

Many thoughts and emotions came through me on my Euston-Manchester train journey. But, two events I saw in the newspaper stuck in my mind that day; the kidnapping of Margaret Hassan, the British humanitarian worker kidnapped in Iraq, and the brain injury suffered by Marc Almond, the lead singer of Soft Cell, in a motorbike accident in London. Once again, I am reminded that the sanctity of good health and living in a stable and civilised environment are more important than anything else.

With hindsight, I should have right from the beginning been much more combative throughout my exchanges with the Dragons. Instead, I had remained subdued and respectful throughout despite my readiness to be combative and feisty. I actually enjoy intelligent debate and particularly didn't want to look meek or intimidated in the show. However, there on the spot whilst still awaiting the response from the remaining dragons, I was constrained from truly expressing my personality for fear of being impolite or arrogant, which might have put off the Dragons who had not yet made up their minds. Also, as a fair-minded Libran, I tend to see the other people's point of view instead of being more aggressive in defending and asserting my own points. Did I regret going on the show? That would depend on the outcome. We shall see. I hoped the programme editor would be kind to me. My failure at the Den did not deter me. If anything I wanted to succeed even more.

## Paris, Smartcards and Hollywood

I was last in Paris in the summer of 1984 – my first trip to continental Europe. Travel then was so much more exciting and a little scary as going about places required a lot more interaction with the locals. I had just passed O-level French and was quite pleased to have used it on my trip there, especially since Parisians were notoriously rude and unhelpful should you not speak French. Wind forward 20 years, in the first week of November, I prepared to visit Paris for its annual 'Cartes' show, the leading tradeshow for the smartcard industry. I would meet Philippe and Alan for the first time, to build up mutual trust and to clarify what's on the table.

Unlike the average traveller, I needed to prepare my travel with extra care and attention. Prior to the trip, I made sure that I adhered to a proper prophylaxis regime at least a week before the trip to prevent an undesirable situation of suffering a spontaneous bleed just before travelling. I bought myself a folding walking stick[132] just in case my vulnerable knee or indeed any lower limb joints got into trouble. Ditto for a new pair of travel trousers which I could easily roll up allowing the knee support to be more readily adjusted. As I couldn't afford taxi fares I bought a light weight rucksack and travel bag with good wheels which minimised stress on my arms and legs whilst ambling along the famous Paris Metro underground system. I used the internet to check out the journey with the least walking distance and fewest staircases between station changes. Just as well, as despite this careful preparation there was simply no option but to use staircases in the Metro. If I had not got one good leg to rely on, those steps would have been an insurmountable barrier to my travel. To minimise the weight I carried, I brought the bare minimum clothing and even calculated the minimum number of bottles of Factor 8 I needed to carry with me for prophylaxis and in case there was a bleed. A letter from my H-doctor about my condition was also prepared in case customs queried the nature of the liquid/powder combination and the needles which accompanied it. Also, I carried a letter from my doctor on my HIV status, in case of prejudice against H. I also made a note of the nearest H-centre in Paris in case of an emergency. A lot of hassle, but it's always better to be safe than sorry. Being incapacitated alone in a busy foreign country wasn't something I wanted to experience!

At last, I arrived in my two star Parisian hotel in a predominately Arab quarter of Paris. I was pleasantly surprised by the accommodating Parisian who so politely gave me directions to my hotel in English, even though I enquired in my schoolboy French. Twenty years earlier, my experience was that the average Parisian would have shrugged their shoulders if you appeared not to speak French well. Resting on the soft bed I was tired but thrilled to have completed the journey uneventfully. I wondered about the next day's meeting with Misters Blot and Finkelstein. What should I say and what would they think of me?

Next morning, I arrived at the massive Villepente exhibition centre near the Charles de Gaulle airport. Streams of black suited businessman and woman were heading into the centre. I probably stood out somewhat as I was dressed in my loose deep green outdoor trousers and a pale lime coloured Berghaus jacket which had plenty of pockets for storing all the tickets and documents for travelling. I should have worn a suit and less combat looking trousers for such an event, but my obsession with wanting to travel light and comfortable made me look rather casual.

I relaxed myself by first inspecting the stands of other businesses, checking out the technologies on offer, and hopefully not finding any products similar to my idea. Eventually, I saw the Audiosmartcard stand and recognised Philippe from the company's website picture. I approached him tentatively and eventually we caught each other's eyes and he knew straightaway it was me. We shaked hands and exchanged the usual pleasantries. I felt comfortable with him straight away. He was friendly and professional, and yet retained a welcoming and approachable manner. One could not help but be impressed by his communication skills. No matter how much one communicates via email or telephone, there is nothing like meeting the person face to face, shaking their hands and getting the feel of their personality. The plan was for Philippe to introduce me to Alan, the mysterious and rather elusive American CEO who a few months earlier had promised to help me to produce my special smartcard! I did not know what he looked like until Philippe spotted him from a distance. My immediate impression was that he looked unusually stylish for a business man, with long straight brown hair to the shoulder, a slender frame dressed in a smart and casual style suit, quite opposite to my preconceived image of a tall and obese middle aged American businessman. He had a longish face, slightly craggy, and come to think about it, looking rather like a less tall version of Iggy Pop, maybe even about the same age. He walked with a swagger that exuded confidence and success. Had he been a younger person, I would have thought him a cocky git, possibly even a gangster. But to have that demeanour for such an older person who was a CEO in a technology business, I instantly thought that he was rather cool. His business card had the title CEO/Inventor which raised my respect for him even further; he was obviously proud of his inventor status, and was not afraid to highlight it. He appeared to be in high demand, always talking to other people and people waiting to speak to him. I caught him just in that lull, shook his hand, introduced myself, thanking him for his kind offer and quickly tried to establish the nature of the offer. Annoyingly, he had another meeting to rush to, but we managed to fit in an appointment for the following day. I felt relieved that I had finally met him but equally frustrated that our first encounter didn't resolve the uncertainties. Hopefully, tomorrow's appointment would finally clarify everything.

I went on to explore the rest of the trade show, meeting with several more industry partners which Philippe had introduced me to. One of these was Wen-Ching, the Chief Marketing Officer of Smartdisplayer technology, a Taiwanese manufacturer and one of the world's leaders in the then nascent E-ink smart display technology (i.e. 'electronic paper' technology as used in popular e-book readers such as Amazon's Kindle). I felt proud that high-tech Western companies were relying on technologies that were designed and manufactured by a Taiwanese company, of people with ethnic Chinese ancestry. We spoke in English as I can't speak

Mandarin and found her to be refreshingly open. She spoke about the potential of the then still fresh E-ink technology, future applications and all that. However, it appeared that their previous display technology products had struggled in the market place. I was surprised by her frankness about the struggles she has gone through in the previous four years of establishing the company. Vast amounts of money were raised from friends and family for the business but it appeared that she and the other key founders of the company had survived on minimum income. As we parted she even implied that we may not meet again next year. Despite the size of investments and achievements that her company had already achieved, I felt sympathy for her and wished her well. There was a sense of camaraderie between us, as entrepreneurs who dared to risk and make great sacrifice for our individual goals. Compared to her, what I had gone through the past year and a bit was just a molehill. I wondered how I would fair in the coming year.

When 3pm came, I waited for Alan, but he was somewhere else. Apparently he was still busily engaged. Eventually, we met up at the VISA stand where he was engaged in a project with them. At last we had an opportunity to talk properly. I could not remember much about our conversation, perhaps because of the paucity of substance, except that there was still ambiguity of how or when he would manufacture the card for me and for what? I worked out that I would have to sort out the 'how' first, and then he would manufacture some sample cards. With the situation a bit clearer, I finished the meeting with greater optimism as he had reiterated his support for my project, and I had had a brief chance to impress my presence in his consciousness. I was no longer just a few email messages, but a real person whom he had to keep his word to.

I went back to see Philippe for the 'how?' solution, which turned out to be a feasibility study which his company could carry out. It would cost about 10,000 euros, a relatively small sum in the world of technology consultations, but sounded a like a small fortune to me. There was a stream of international visitors to Philippe's stand, and I was introduced to them, who all turned out to be very friendly. I had not expected business people to be so open. The event had felt like a friendly academic conference with everyone wanting to openly share their knowledge and contacts. I started to feel like part of the family. One of the American's whom I was introduced to started to talk about Alan. I was very keen to hear more, as prior to that, there was very little I knew about him other than that he had previously marketed a credit card with a built in magnifying glass. Apparently, he had been a restaurateur in New York and Los Angeles before becoming an entrepreneur in the credit card industry via his 'lenscard' invention. I momentarily felt an extra sense of deja-vu about my chance encounter with him, as I too was involved in the catering business before, except I was just a shop assistant in my parents' small Fish & Chips shop. More excitingly, he had also dabbled in the movie business, having been an associate producer of 'The Two Jakes', the sequel to Jack Nicholson's famous 1974 film 'Chinatown'. Alan and Jack Nicholson were mates and had been business partners in the popular Monkey Bar club in Los Angeles. Wow, that must be how Alan got his cool. Jack Nicholson is just about the coolest actors there ever was, and despite his age still carries with him the gravitas of a legendary star. I have not previously idolised anyone except great scientists such as Newton, and Einstein, so I felt embarrassed to admit that on hearing Alan's connection to Jack Nicolson I felt a bit awed by this rather irrelevant and indirect connection of a world famous movie star. I imagined that the great Jack Nicholson may well have heard of my Smart card alarm idea via Alan. And, if I were to collaborate with Alan successfully, I might one day even be invited to meet his Hollywood pals.

That evening I felt relaxed; my job at Paris was done for now. I gave myself a prophylactic shot. Tomorrow I would spend the last full day sightseeing and visit my uncle and his family. He was running a small Chinese traiteur ('takeaway') on Rue de Rivoli, a few steps from the Musee du Louvre. Twenty years earlier, he had arrived in Paris as an illegal immigrant. Now he was running his own business, and to my great surprise spoke what to me sounded like impressively fluent French!

It was great to be back in England, coming out of Leeds Bradford Airport, breathing in the cool fresh air of the North Yorkshire countryside as I made my way home in my cute little French car, a Citroen C3. This journey had given me back the confidence to travel again; perhaps next time it would be America, and the world if necessary.

As a matter of interest, a few years later I discovered a particularly interesting story of Finkelstein's lenscard invention and the challenge of being an inventor: Nicholson and Alan were dining at Abertone at Aspen, and as the waiter arrived with the bill, Nicholson pulled out his credit card and a plastic magnifying glass[133].

'What are you doing?' Finkelstein asked
"I am trying to see the bill..." Nicholson said.

This was an eureka moment for Finkelstein. Immediately he thought why not combine the two? A credit card with a built-in magnifying lens is certain to fulfil a market gap. As a co-owner of the ritzy IndoChine restaurant on Beverly Boulevard, and previous owner of the infamous Monkey Bar with Nicholson in LA, Finkelstein was not like your typical 'poor & lonely' inventor. Given his means and contacts, it shouldn't be too difficult to realise his idea.

"Everybody thinks, 'Wow! what a simple idea" Finkelstein said. "People think all you need is cut out little holes and glued in little lenses."[134]

But it took seven years, investment of several million dollars, series of patents, working with several manufacturers and much hard persuasion before he got his unique credit card to market. The deal was made with Chase Manhattan Bank which launched the 'LensCard' at a party attended by Finkelstein's celebrity friends Nicholson, Michael Douglas, Sean Penn and others.

One point in the above story which really struck me is that it took SEVEN years to go from conception to launch. This is as long as doing two more PhDs consecutively. I hope I could achieve my goal in far shorter time as the thought of struggling to 2010 just to reach launch date would be unbearable[135].

## Good news, bad news

I left Paris excited, but also at the same time anxious about the challenges ahead; where would I get funding to start the feasibility study? How could I extend my patent application as the 12 months deadline from filing drew ever closer? When would Dragons' Den be on? Perhaps the publicity could bring me the investment which I now desperately needed.

Having connection with the local University brings several advantages. There were frequent talks and seminars from speakers on various business matters. The most relevant for me were talks about funding opportunities for innovative technologies. One of these funds was the York Innovation Fund, run by a joint venture between the University of York and the local council to encourage technology start-ups. A few months earlier, my qualification for this fund would have been premature, but now I believed I had a chance. Indeed, thanks to my persevering contact with one of the business promoters, Carolyn, and the demonstration of my commitment, she became my champion. When one gets a champion on your side, the chance of winning is much greater. After the usual application form and the compulsory business plan which I had already prepared for the Dragons' den application, I was finally awarded a £10K York Innovation award. This was not a grant, but a part equity and loan deal with a generous preferential payment term. This funding support meant I could a start a new company to drive my vision and dream forward. After some thoughts on branding, in December 2004, ProActiv eCard Ltd ('ProActiveCard') was incorporated.

I have to admit that when I began my entrepreneurial quest to develop my inventions, I did not imagine myself getting into debt. The sacrifice of going part time and then full time dedicated to a new venture with zero pay and mounting costs seemed a big enough risk already. However, there was no regret whatsoever. Since leaving my part time job, I had had an exciting and fulfilling start to my full time entrepreneurial quest. It appears that once I start something my nature is to see it through all the way. But 2004 hadn't been just challenging business endeavours. There were also ugly new developments for the H-community. By pure coincidence, on the same September day that I received the ex-gratia compensation for my HCV infection I received another letter from the Department of Health warning of the possible transmission of variant Creutzfeldt Jacob Disease (or vCJD which is the human form of mad cow disease BSE) from blood products. A blood donor whose plasma was used in the production of F8 had developed symptoms of vCJD six months after donating blood in 1996. All H-patients who had received treatments between 1980 and 2001 were classed as at risk of being infected, and with the potential of developing the deadly brain disease. This was really the ultimate sweet-bitter moment. I was for once extremely angry and pissed off of yet ANOTHER blood contamination fright! However, my feelings did calm down fairly quickly though. Perhaps this was just my optimistic nature of coping, by shutting myself off from the terrible thought of the consequences. After all, I had beaten the odds of not having been infected by HIV, and apparently staying healthy with HCV. Maybe I am lucky and have a particularly strong immune system or is less susceptible to infectious diseases. The mathematical part my brain wishfully assured me that the probability of developing vCJD must be minuscule since Mad Cow disease (BSE) was known to be in the UK food chain from 1980 to 1998. Therefore, it is probably safe to assume that a very high proportion of the British population must have been exposed to BSE which was believed to be a major cause of vCJD, and yet so far there were less than 150 cases of the deadly disease.

A few days before Christmas I had my quarterly appointment with my H-consultant to discuss this new development. He offered me the choice to find out if I had been one of the 400 people exposed to the 200 batches of the tainted Factor 8 which was contaminated by the donated blood of the British donor who subsequently developed vCJD. As there was no test to detect vCJD, and I didn't want to dwell on it, there was no point in knowing my situation and so I chose not to know. But unbelievably, the doctor misheard me and said 'yes'. I had been exposed! Flipping heck, how could he mishear me so badly?[136]! This misunderstanding could have been devastating for some people. I tried not to think about it much although inevitably I did. Of course it would have been better if the doctor had confirmed positive news, but to be confronted with a negative certainty shocked me somewhat, even though I kept a polite straight face. I don't blame my excellent doctor for having misheard me though. It was probably my fault as the stressful situation probably made me mumbled through the reply. Did this change my outlook given that I now knew for certain of my exposure to a 'high risk' batch? At the time I thought not because I had already normalised my acceptance that the Factor 8 blood product carried an inherent risk of contamination, whether of known or unknown nature. Even if I was told the opposite news, I would have still assumed an inherent contamination risk as Factor 8 was manufactured from a pool of up to 30,000 blood donors. My reaction was also tempered by my HCV status which I considered as presenting a much more eminent concern. Therefore, at the time I did not dwell on it too much. But later – news and changes to my personal circumstances brought much greater anxiety as I will reveal later. There was also good news in the treatment of H that year. From the beginning of the year, adult H patients, including myself, were finally offered Recombinate Factor 8 which was the first generation of genetically manufactured (or 'GM') treatment, thus virtually eliminating all the inherent human blood plasma infection risks. There was still a minuscule risk though, as this first generation GM treatment still required the use of human derived albumin to act as a stabilising agent, as well as traces of mouse protein[137] in the manufacturing process. Perhaps because of the terrible vCJD development, even this advanced treatment was further upgraded by the end of the year when since then almost all H in the UK have been given the so called 3rd generation GM Factor 8 which doesn't have any human or animal proteins in the final product. Although my blood had already been tainted, there was a strong comforting sense of relief knowing that all my future treatment for H would be completely free of the risk of catching HIV, vCJD or any other as yet undiagnosed human blood disease.

## Annus Mirabilis?

The start of 2005 was going to be expensive since the 12 months deadlines for the Reminder smart card and iiCap inventions were approaching fast. The York Innovation fund had solved one of my immediate concerns; I could at least maintain the reminder smart card patent application. The options when faced with a patent approaching its 12 months grace period are; 1) abandon it; 2) proceed with the UK application only. The patent must now be supplemented with a full set of legal 'claims' derived from the description given in the application (quite expensive). Furthermore, it is then necessary to apply for search report and a request for examination by the UK patent office (not expensive); 3) proceed with entering into the national stages of as many countries as possible (very expensive). And finally 4) enter into the so-called Patent Cooperation Treaty or 'PCT' application with the World Intellectual Property Organisation office (fairly expensive). The PCT application is sometimes erroneously referred to as a worldwide patent. There is no such thing as a world patent, the PCT application is basically a pending international patent application which gives the inventor a further 18 months extension before they have to decide whether to pursue the national stages.

There are pros and cons for each of the options. Option 1 in the shortest term is obviously the most economical, but potentially the most expensive as a lack of a patent application could deter investors looking for barriers to entry or intending to use the patent as collateral should it be granted. Option 2 is the next 'cheaper' route with the advantage that the patent application is maintained for at least the home country.

Option 3) is too expensive for most start-ups especially if they don't have any turnover or significant investment. Hence, at the time, I thought option 4) was strategically the best option because the extra time gained meant more time to develop the business and left open the option to subsequently apply for patents in other valuable territories. The fees for the PCT application alone cost about £2000. The lawyer fees for filing the complicated PCT form cost an additional £1000, and when other services are included the bill was over £5K. A sizeable chunk of my recent loan gone already! I also made the PCT application for the iiCap invention, although I filled in the application myself to save money which had to come out of my savings.

If I were to make further progress, there would need to be more investment than my limited resources afforded. My networking the previous year had alerted me to the availability of a Research and Development grant from the Department of Trade and Industry. Through the help and advice of Businesslink Yorkshire, I submitted the application in the first week of January. I also arranged to see Mr Finkelstein in Los Angeles in February with the purpose of pinning down the exact commitment which he had promised. After months of uncertainty, the BBC finally confirmed that the first series of Dragons' Den would be broadcast in January, and some clips of me would be shown. This was a big relief as the associate producer told me earlier that not all filmed contestants would be aired. It would have been utterly disappointing and despicable if after all that effort, expectation and humiliation, one did not at least get some airtime exposure.

All the while, my increasingly troublesome knee was never far off my mind as I continued to cope with the pain and limitation of living with H. Unusually, one of the H-specialist physios, Thuvia was also trained in acupuncture and acupressure. Initially anxious, I requested a trial of acupressure first as it only involved applying finger pressure at specific points – without needles! The treatment obviously would not cure my knee, as it was far too damaged. But it might help to reduce the pain, improve circulation to the joint and reduce the swelling. My first course of acupressure involved application of pressure around my knee, foot, hand, and even at a point on my earlobe. As much as I wished for some miraculous effect, it didn't do much for me. But, on my second course Thuvia warned me about a particularly powerful acupressure point on the side of my knee which is known to have a particularly noticeable effect. Incredibly, almost as soon as the treatment was completed, I felt a strong sensation of fullness in my knee, a sort of increased pressure within it. Unlike a bleed which disables the knee and causes painful swelling, this effect did not cause that, but a positive feeling like there was a higher flow of circulation to my knee. The effect was felt for about 2 days, and at the end of it, the knee looked and felt firmer and tighter, with less puffiness than before. I didn't have any doubt that the feeling after the treatment was very real, but could it be just coincidental or purely a placebo effect? Since the effect was very specific and strong, I should be able to conclude on my next treatment if the cause and effect was real or not.

At last, my slot on Dragons' Den was to be shown on the first day of February. The program had already established itself as an unexpected hit with over 5 million viewers, and a must watch for almost anyone interested in business and reality TV. I hoped the editor would give me a long slot and cut out all the embarrassing bits. From the preceding five episodes, it appeared that the latter was a must for the program as I had already caught preview clips of myself in several compromising situations. The format had 3 slot lengths; 1) a long one where much of the contestants presentations and their battle with the dragons are broadcasted, 2) a short clip of about half a minute that summarises the encounter and barely gives much information about the contestants or their products, and 3) a few seconds snippet clip that just barely shows the contestants walking up the stairs! And as transpired after watching all the programs, some contestants did not get any airing at all; one of whom included the 'door-safe' inventor which I mentioned in the previous chapter. He must have been really pissed off! One of the producers had already warned me that I did not get the long clip, so I was really anxious that my clip would at least provide a brief demonstration of the product idea. I did not walk away with an investment from the Dragons, but hopefully the publicity may just attract others.

I had never been on TV before, so it was with great anticipation that I watched the program that day. It turned out to be a 30 seconds clip, packed with catch phrases and mishaps. Whilst being asked by Doug Richards to demonstrate the Smart card alarm, I replied embarrassingly that "The battery has gone flat in this one". Simon Woodruff thought the smart card alarm would be annoying as it emits the alarm sound. He said to me "If you are ever successful, you are going to be the most unpopular man in England". So far, so good… it was sort of funny but it hadn't shown me or the product in a seriously bad or embarrassing way. In fact, the most entertaining editing was reserved for the last few seconds as Woodruff further laid into my idea by exclaiming animatedly "Have you seen the new thing the National Health Service has sent you? This thing is driving me crazy; I can't find the stop (alarm) button on it". Evan Davis interjected in the edits saying "But Kin has the last word", to which was a cut in which I replied "But you MUST be an idiot if you can't find the button!" In the background was Racheal Elnaugh laughing hysterically. And that was the end of my clip. At this point, I also burst out laughing, delighted that it ended with one of the dragons being made fun of. For the first time in the series, this was probably the only clip where the laughs were on the dragon instead of the contestant. That response at that moment was a total surprise for me too as it was said over three months earlier. Actually, I had not intended to be rude to anyone. I still recalled vividly at Simon Woodruff's totally contrived animated objection, so I thought to myself as I looked at my card which has only one button, that whoever that could not figure out how to shut the alarm must therefore be an idiot. But, in the filming, my mind was stuck for a more polite phrase, so I simply said what I thought! And I'm glad I did, as the show at least gave me back one highlight that showed my fighting spirit.

So, was it worth getting on the Den and would I recommend it? The answer is a categorical 'yes' provided your edited clip is not entirely negative. If one's business is at an early stage like mine was, the application process can instil positive pressure and discipline that allows more work to be done. The free publicity can be massive, which can attract other investors. Unfortunately for me, my clip was brief, but it was long enough to catch the attention of a CEO from a North East blind people association. She alerted me to the needs of people in vulnerable sections of the community with memory lapses. This knowledge and her support further enhanced the market argument. The program at the time had also established a particularly fearsome reputation; so many people who saw me in the program automatically granted me the 'brave' accolade. My edited appearance on the program definitely enhanced my seriousness as a determined entrepreneur, and in the eyes of some of my friends and acquaintances me being on TV was seen as being quite cool. In fact, to my much embarrassment, some of my friends used this as chat up line in pubs, always the useful wingman! So there was no regret of having been to the den, but with hindsight I wish I'd tried to be less dignified and had tried harder to appease the dragons. It would most probably not have changed their minds, but the editor may have given my presentation more exposure.

On the other hand, if you or your product is shown in the most negative light only, then there is very little to recommend the program for. The chance of obtaining further investment could be even harder after such an appearance, and that is before one considers the humiliation of being so cruelly depicted on TV. For example, the lady contestant with the toy idea who was filmed on the same date as me, was shown with tears on the edited clip, and was then comprehensively demolished by Rachel Elnaugh as unworthy of being a businesswomen for being overemotional. I was lucky that my clip turned out quite kind despite being rejected by the Dragons. The televised bits showed fun bits rather than any substantive arguments against the reminder smart card. Just because the Dragons are experienced and successful entrepreneurs, it does not make them experts in the industry sector in which they know very little of. There have been many classic examples where they have been wrong. Because the dragons have to make their investment decisions on the spot without proper due diligence, one may notice that almost all their investments are on the 'sure thing'; businesses that were already trading profitably or have firm and significant contracts in place. Another questionable aspect of the program at the time was the apparent lack of due diligence on the dragons themselves. Whilst Rachel Elnaugh acted as a wealthy investor on the show, her company was in such serious financial problems that it eventually needed bailing out from the other Dragons! I felt quite angry about that as we contestants were subjected to mountain loads of due diligence, but the BBC had apparently not even bothered to check her company accounts. This meant some entrepreneurs could have lost potential investments from a more serious investor. Simon Woodruff was another waste of a slot as he never made a single investment – but I can't really blame the BBC for that!

Although the primary point of my trip to the US was to establish a firmer relationship between the elusive Mr Finkelstein and myself, it was also a bit of a holiday, and an opportunity to see my great aunt and cousins in sunny California. The transatlantic flight to America was very comfortable as the cabin was barely occupied. I used some of the spare time to prepare my meeting as well as thinking up new card ideas which might impress the established card tech CEO. To appear more professional, the preparation involved a consultation with a commercial lawyer a week earlier who advised me on the terms of any future working relations. The post 9-11 strict security was noticeable with shoes removed in custom, close contact body searches, and stern immigration officers demanding fingerprinting and iris scanning. Fortunately, once inside the country the security for the internal flight between Atlanta and LA was much more relaxed. This six hours journey from the East to the West coast was an impressive reminder of the vastness of America. Not familiar with US geography, I had no idea of where we were at any instant of the flight, but even at 30,000 feet there was amazing scenery, including an apparently smouldering volcano which had somehow imprinted firmly in my mind. Alas, I arrived at the legendary LAX airport, first time in the city of angels and stars. I chose to stay in a hotel rather than with my distant relative. I prefer the privacy because of the awkwardness of sometimes needing to treat myself, or when I needed rest instead of being sociable. I understand my relatives knew about my condition from the worries and pain of my youth, but they didn't know much beyond that; besides I don't like to impose on people.

It was great to arrive at the 3 star hotel. Like they say about America, everything was bigger and cheaper; my ground floor hotel room was massive, probably at least five times larger than the Paris hotel room I stayed in a few months earlier, and costing about the same. I booked the stay for 12 nights, to make sure there was sufficient opportunity to complete my mission to get Mr Finkelstein's commitment. Just as well, as I came down with a severe cold on my arrival and stayed mostly lying down in the rather chilly room where the outside weather was abnormally cold and drearily wet with periodic bouts of heavy rain. Looks like I had brought with me the Manchester weather! As usual on a foreign trip, I was diligent on my prophylaxis treatment as it would be disastrous to have a major bleed. However, it seemed to me that the weather didn't help, and there was a constant 'just under control' pain in my ankles which made walking harder than usual. I got my first appointment with Alan on my third full day in LA. Before then, I used the free hotel shuttle bus to visit the famous Santa Monica and Venice beaches. To my great disappointment, there were no Baywatch babes. Unfortunately, the February weather was not the best time to enjoy the beach!

Alan's office was on Wilshire Boulevard a few miles up from Santa Monica beach. The hotel's free shuttle bus service to Santa Monica was an excellent gateway to the business district of LA, where I could then get a much cheaper taxi ride to my final destination. The journey up this road seemed to go on and on. I guessed it was just one of those psychological effects where the brain is busily processing new scenery. Finally, I was dropped off near the building. The Latino taxi driver was unsure, but it was close by. As I looked up the street,

I was pleasantly surprised and delighted to see a very familiar face looking down on me. There were a series of large posters waving on the tall lampposts on each side of the road, of a wild grey-haired, slightly portly middle-age man precariously balanced on a pedal bike. Oh wow, it was my childhood hero, Albert Einstein! There was an exhibition about Einstein in the Skirball Cultural centre on the outskirts of LA. Of course, 2005 was the centenary anniversary of 'annus mirabilis', the miracle year of 1905 when Einstein theoretically proved the existence of atoms, the explanation of light as a quantum particle, and of course the famous Special theory of Relativity where he uncovered the reality of space contraction, time dilation and, of course, the world's most famous equation $E=mc^2$. Apparently it was the most comprehensive exhibition ever mounted on the life and legacy of Einstein. This was one show which I would definitely not miss.

Einstein aside, I entered the tall black windowed building with much trepidation, elevator up to the 21st floor and into suite 2150. At last there was going to be some properly allocated time to discuss business. I was greeted by Nick and then Thep who emerged from a meeting. It was wonderful to meet them in person and to put a face to their American voices. Although they were not the decision makers, they had been instrumental in convincing Alan of the benefits of the Smart card alarm. I could hear Alan engaged in an animated phone conservation in his office. I noticed the pretty receptionist was also an Asian (probably Chinese) American. Maybe Alan really liked Chinese people; maybe he would make me his Chinese British partner. Finally, he was ready and I entered his sumptuous room with the spectacular California hills view in the backdrop. Although smiling, he looked serious and quite hurried as though there was some urgent development from the previous phone call. We exchanged the usual pleasantries, and presented him with a small box of premium English tea bags, a quintessential English gift I suppose. I quickly got into the business side, asking about his experience raising finance, patents, and how he could develop the card for me. From the meeting there were three things he said which stuck in my mind: "Patent is a game; useful for investors, but cost a fortune", that "I would be successful with my determination" and that he would help me build some cards as a gift for me provided I have the design specifications. The latter statement was a marginally more concise commitment than I'd had before, and it felt more reassuring that I'd heard it directly from the boss himself. However, it was all just spoken words. I really wanted more detail, but it felt inappropriate to demand a legally binding commitment. The Smart card alarm was clearly still in the embryonic stage, and my expectation that he would just make the prototype cards from scratch was just wishful thinking.

As we moved to the kitchen to make a drink, my memory of the swagger in his walking style was not exaggerated. There was almost a gangsta-like stroll in his walk, although I suspect it was an unnatural gait developed over a period of time mingling with his celebrity pals. Alan then proceeded to show me his company's latest card innovation; a VISA credit card with a built in LED light. It was a nice demonstration of a fully embedded card featuring power consuming electronics, and a nice gimmicky marketing tool for a credit card. Perhaps I was too focussed with my own idea; I didn't show much interest of his demonstration, which in hindsight was somewhat impolite. As he had a connection with the entertainment industry, I showed him my mini clips on Dragons' Den for a bit of extra bonding. He invited me for dinner the following week, so hopefully we could build further rapport then. Thep printed out the directions to the Skirball Jewish cultural centre, and with our first meeting finished, I left the building to emerge into the familiar and jovial presence of Einstein. Feeling more relaxed, I looked forward to a weekend of rekindling union with old and new relatives, as well as my childhood hero. There was some pain and stiffness in my left hip by the time I got back to the hotel. This was an unfamiliar feeling as it was awhile since I'd had a hip bleed. Not taking any chances, I sorted it out with a dose of Factor 8 before it got worse.

By Saturday, the LA weather had returned to its norm for that time of the year, mild and not cold and damp anymore. Mary and her mother who is my great aunt picked me up from my hotel. I'd not seen them for at least two decades. They last visited my parents in Oldham, and recalled how they'd a relaxing stay because the air and climate near the Pennine moor was so refreshing! Perhaps not surprising as smog was endemic in LA at that time. They were part of the great Chinese diaspora who first left China for Hong Kong a decade before my dad, who then subsequently migrated to the US. Despite all the glamour of America, and California especially, I felt fortunate that my dad had chosen to emigrate to England instead – the private healthcare system in the US could have made my family's life extremely precarious. My great aunt and late great uncle were devout Christians who had influenced my parents when they were living in Kowloon back in the old days. Despite the years apart, they were instantly recognisable as I approached them in the hotel's car park. It's always nice to see old friends and relatives, but there's always some awkwardness about how to greet them. Despite years of living in the West where people are much more touchier, the Chinese of my generation and earlier had retreated to an almost no-touch policy. I suppose it's a kind of repression. Shaking hands seemed too formal and distant, so often I simply gestured and proceeded with the usual greeting conversation quickly. As it turned out, Mary extended her hand to me which initiated my greeting hand, and as we grasped each other's hand she followed up quickly with a light hug. Immediately, I felt the burden of the greeting protocol lifted and I proceeded to copy her style on my great aunt. We had a nice time catching up, visiting her middle class home on Beverly Drive, and as is the Chinese way, went to a Chinese restaurant in LA's downtown Chinatown for a meal. Although the restaurant was busy, interestingly, the LA's Chinatown was eerily quiet and darkly lit in the evening, almost like the quite scene from Jack Nicholson's famous 1974 film Chinatown! Apparently, there'd been some gang warfare recently which had made Chinatown a less than favourable place. It seemed many parts of LA were not particularly desirable in the evening as I was warned to avoid even the famous Venice beach!

Next day, Mary took us to the Skirball cultural center to see the Einstein exhibition. En routes we drove on the famous Mulholland Drive road. I recognised the road's name because it featured on many films and documentaries, even though I could not name any at the time. I felt so lucky and excited to be in LA. To have the unexpected opportunity to attend the most comprehensive exhibition about Einstein. I was not disappointed. There were the famous exam results which showed him failing French; the Nobel medal which appeared to have some of its top part missing (perhaps from the thousands of people who had touched or even scraped from it over the years before it was securely boxed); there were some of his original notes on Special and General Relativity where Einstein showed his famous derivations about light, matter, energy, and of course the intimate relationship between space and time. I came on business to see a Mr Finkelstein, and yet my fate had also brought me to see my all-time hero, Mr Einstein! Perhaps, 2005 would be my own mini Annus Mirabilis?

It was a fantastic day, but by evening, I could barely walk as both of my ankles were in a lot of discomfort. The pain on the left ankle had definitely deteriorated to the point where I could not comfort myself with the false belief that it was just merely arthritic. It was clearly a bleed in the joint, and possibly in the muscle tissues above the heel area. I forced myself up from my bed at midnight to promptly administer a F8 dose. It was fortunate that I did not procrastinate further that evening. From previous experience of bleeds in that region, it could have debilitated me for several days or more.

The dinner with Alan was cancelled at the last minute, which pissed me off. Once again he appeared to be so hyper-busy. I was looking forward to the opportunity to get to know each other better. Of course, my additional motivation was to increase our bonding a little more by demonstrating my commitment and determination. This way, it would have been harder for him to change his mind. As I was to find out later, he was probably distracted by the Hollywood actress Rebecca de Mornay whom he was dating in 2005. Instead, Thep and Nick took me for lunch instead; which nearly cost Nick his life as he was almost run over by a speeding car on a pedestrian crossing. Fortunately it just missed him. The meal and company were nice, but I left frustrated not having anything concrete to take home.

With my remaining few days, I visited more tourist attractions, the usual places, including Hollywood, Beverly Hills and Long Beach's Aquarium of Pacific. I had started to fall for California; the weather was great, the sea which I had always loved, the fresh food, the proximity to great tech companies and Universities, along with sharing time with relatives had all made me feel quite at home. Perhaps if my business became a success, I would have more excuses to visit more often. Even better, I might even get a Californian girlfriend. On my last full day in LA, I returned to Alan's office. I was determined to get something solid from my 6000 mile trip. I had prepared a letter for Alan to sign. Borrowing Thep's computer, and using the company's formal letterhead, I typed the letter on Alan's behalf which stated all the support he said he would do. Fortunately, he did not object and signed the letter. Great! Meanwhile, Thep passed me a note from the pretty receptionist. She wanted a date with me! Wow! American girls sure don't take no prisoners. It must have been my sexy 'British' accent, and what a blissful way to end my trip!

I had to decline, unfortunately. I made the genuine excuse that I had a prior engagement with Mary and her husband. Actually I would have rescheduled or even cancelled it if it weren't for the other reason that I had been really struggling to walk normally over the past few days due to the pains in my ankles and my tired knee. I did not have the confidence to go on a first date walking with a severe limp, making up lies about the real reason, trying to hide my pain or discomfort, with the very real possibility that it might get significantly worse. One of the really frustrating things about H is that it frequently takes out the spontaneity of normal living.

## Chinese American vs. British Chinese

Perhaps this is a good place to interject my comparative view between the British and American Chinese experience. The Chinese American story tends to be more diverse than the UK one where most of the earlier immigrants worked in the catering business. My late great uncle, who also worked as a seaman in the generation before my dad, had chosen to go to the US rather than the UK. There, he and my great aunt worked as housekeepers. Unlike in the UK, these were popular jobs for these early Chinese immigrants with limited English language skills. Although not glamorous, the American economy at the time could offer good financial prospects for almost anyone who worked hard. Their daughter Mary with her Chinese immigrant husband ran a successful garage, while a second cousin was busily occupied in the Oscar presentation that weekend. Today, the Asian American is one of the most successful ethnic groups in the USA, disproportionately overrepresented in professional jobs, as well as start-ups in science and tech related industries. When I was growing up I was inspired by many great people irrespective of their race or nationality, but it would be false to pretend that one isn't more inspired to see a person who shares much of your identity achieving global eminence. In this sense, America has been the home of many notable figures with Chinese heritage. Personally, the following people have inspired me in my youth: the first Chinese American Nobel prize winners CN Yang and TD Lee in physics; Qian Xuesen, cofounder of the Jet Propulsion Laboratory; Dr An Wang, pioneering computer entrepreneur and inventor[138]; Steve Chen, principal designer of Cray's most successful supercomputers; Dr David Ho, the Aids researcher who pioneered the successful cocktail approach to fight HIV, Jen-Hsun Huang of Nvidia Corp, Charles Wang of Computer Associates; Jerry Yang of Yahoo!; Yo Yo Ma; I M Pei; Steven Chu, Nobel Prize winning physicist in superconductivity and former US secretary of energy. The list goes on!

On the other hand, growing up in Britain, there were almost zero famous British Chinese characters as role models. Burt Kwouk and David Yip of 'The Chinese detective' were the only noticeable characters with some public presence, although only

David Yip didn't play the stereotypical role. One noticeable exception is Kuen (Charles) Kao, the Hong Kong born and educated physicist who did most of his great work on the theory and application of optical fibre technology in England. When he won the Nobel Prize in 2007, he was claimed by China, Hong Kong, Britain and America as one of their citizens!

Today, there are a few more notable British figures with Chinese heritage; Jimmy Choo, Anna Chen, Linda Yueh, Gok Wan, James Wong, as well as the many unsung scientists and academics that are increasingly making significant contributions to British life and culture. There are obvious differences that can explain the disparity in the top level achievements of American and British Chinese. Such as the much larger population size of the former and the high concentration of Chinese Americans in the most demographically dynamic areas of the East and West coasts of America. The quality and quantity of top US Universities and the preeminence of the US IT sector has also attracted the best in the world to study there and the creation of start-up companies, especially those of Taiwanese origin. Compared to the UK, I believe the latter circumstantial factors have been especially lacking in the UK in the last century. However, the beginning of this century has seen a dramatic rise of technology start-ups in the UK, and an absolute growth of highly skilled Chinese students and professionals arriving in the UK; thus I predict the balance of British Chinese contribution in innovative areas is likely to improve significantly in the near future. Just over the past 5 years alone, I've noticed a significant rise of Chinese and British Chinese holding higher academic positions, whereas there were hardly any during the 80's and 90's. And, according to an estimate by Financial Times technology correspondent Marcus Gibson, "at least half of the high potential University spinout companies have been founded or co-founded by a Chinese- or Asian-born graduate scientists"[139]. Hence, I am optimistic for the future prominence of the British Chinese community.

## Patent is not everything

Prior to my US trip, I received a professional search report of the Smart card alarm patent. It was conducted by the DTI grant body for the purpose of evaluating my grant application. Given what I knew already, the response was unsurprising; several prior arts were cited, and my patent was deemed unlikely to be granted. This was an obvious set back and a barrier to the success of the application. However, I was prepared for this from the encounters of the previous year. I would need to convince the authority that a patent search is not always conclusive. Search reports don't usually go into crucial details and methodologies which are often misunderstood or neglected completely. The other point is that even though intellectual property such as patents are potentially very useful, they are not everything when it comes to developing sophisticated hardware. The reality is that all high tech products have a plethora of patented components and technologies within them. One is likely to be making use of other people's patented technology in one way or another; licensing IP is an accepted cost of all such products. The barrier to market is not just about owning a patent, but about being able to create a great product with decent commercialisation strategy and execution. In creating a great product, there is likely to be new valuable IP that doesn't always need to be patentable, such as proprietary codes or algorithms that would be better kept secret rather than being published. Could these arguments convince the awarding committee that owning a patent is not everything? There was no laurel to rest on, but hopefully my trip to America, together with my perceived brave Dragons Den attempt and all the relevant support letters would swing the decision in my favour. I'd also developed an excellent rapport with the grant advisor and project administrator Peter Griffiths who had wholeheartedly supported me in the final review process.

The result was announced on the 18 March 2005. The call from Peter finally came in late afternoon. The news was brilliant; I had got it! Finally, I felt like all my previous efforts had been justified, and the development of the Smart card alarm could finally begin. The grant was worth £12,500 over an eight month period to complete a specification and feasibility study of the innovative card. This was a matching grant, which meant I had to show my company could supply a similar amount to complete the £25K project. This was not a small sum to fill, especially as I had already lived on my savings for the past 6 months!

The grant accelerated the development process, but it didn't give much relief to the burn rate of my personal savings; I had to seek as much financial help as possible to survive. Fortunately now that I had a small income as an employee of my own company, my accountant alerted me to Gordon Brown's working tax credit allowance for the low paid. This supplemented my income by a few hundred pounds per month and made it possible to reduce the burn rate somewhat. Gordon Brown's working tax credit has been criticised for subsidising rich companies' payroll, but it may also have unwittingly encouraged many new SMEs and innovative entrepreneurs to grow!

Compared to the costs usually associated with developing new technology involving both hardware and software development, the cost of the project was trivial. This was made possible because Audiosmartcard subsidised their services on the basis of a joint venture-like agreement. They offered the feasibility study for a relatively low cost of 11,000

Euros, subject to them being the preferred supplier for future mass orders. It was exciting to be finally directing my first technology project, but at the time it was intimidating to have to frequently deal with new technical jargon, confusing terms and conditions, and other lengthy legal documents. There was also some complicated legalese about ownership of intellectual property. Careful reading appeared to indicate that Audiosmartcard retained the IP of the feasibility study even though it was my idea and money which initiated the study. Proper legal clarification could have dragged on for months and delayed the project. Fortunately, Philippe Blot, who had been very helpful throughout, quickly negotiated a joint IP agreement which I felt comfortable with.

## An interlude of normality

The realisation of the modest grant meant there was at least a small income for the rest of the year, and a clear focus in achieving the goal of completing the project of creating the world's first Smart card alarm. It felt so much more productive to be engaged in actual product development work, rather than chasing after funding or investment. There was a sense of normality returning and an expectation of reaching the final goal of creating a useful and profitable product.

I went to Paris a few more times to supervise the project to ensure it achieved as much as I had wanted. By that time, Audiosmartcard had been taken over by a large smart card company in an impressive black windowed office tower located in a posh suburb of Paris. I was introduced by Philippe to the company's young technical lead Didier, a brilliant technical wizard of African heritage who headed my project, and the company's founders CTO and CEO. Didier and Philippe both had wonderfully open and easy approachable manners, whereas I felt the CTO and CEO rather offish and distant in a stereotypically French way. If my first point of contact were them, I doubt I would have gotten this far. There, I started to discover the nature of technology start-up companies. Philippe informally told me the company's history; he casually referred Audiosmartcard as a start-up company even though the company was over 5 years old, and told me the company had never been profitable in all those years! As an inexperienced lone inventor seeking to grow my own technology company, I was quite easily impressed by any company that could employ more than a handful of full time staff. But the reality was that many technology start-ups require sustained investment for long periods before they can expect to turn a profit. Often, many of these companies fail. In the case of Philippe's company, it was funded by LVMH, France's world famous luxury goods group which owns numerous famous brands such as Louis Vuitton and Christian Dior. I was quite surprised by Philippe's matter-of-fact frankness, and reflected on the meaning behind his words: many start-ups can exist for many years even when they are unprofitable, so long as they retain the potential to become big, and maintain a palatable exit value to their investors.

During the development phase, I continued to network and look for investment, including sending an investment pitch to Bernard Arnault, France's richest man and majority owner of LVMH. I thought naively that with an introduction from Philippe there would be a good chance of eliciting a response from him, as my Smart card invention may well be the killer application which could also benefit his investment in Audiosmartcard. There was no response though, nor any success with other investment pitches. Basically, it is extremely difficult to find investors during the development phase unless one is pursuing a particularly hot topic. Investors tend to go with what is trending. For example, the current hot topics are social media businesses and renewable technologies. However, there was one particularly fruitful encounter during this time. I was invited to a barbeque party by a Chinese friend who was married to a German professor of Chemistry. The party was held in the garden of a grand house previously lived in by a bishop right next to the spectacular York Minster on one side, and overlooked by the city's historic Roman wall. There was a contingent of Chinese graduate student guests in the party. Over the beautiful scenery and aromatic smell of our barbequed foods, tourists from around the world doing the picturesque wall walk looked on curiously (or perhaps jealously) at us as we enjoyed and chatted under the beautiful mild July Sun. At the party, I met Heng Liu, a PhD student from China who was working on embedded systems in the Electronic department. He was impressed by my endeavours with the card project and we subsequently arranged to meet again. This was my opportunity to add to the team someone with excellent programming and hardware skills. Without much hesitation, Heng became the company's first partner and my company ProActiv eCard Ltd was no longer a one-man band.

As usual, the project hit a few snags and took longer than expected. There was another higher priority project; ironically it was Finkelstein's latest card project which was delaying my project's development time. The feasibility study was eventually completed at the end of the year. Disappointingly it did not entail the creation of an actual smart card size prototype. That would have involved additional manufacturing at considerably greater cost. Rather, the feasibility provided only the necessary component specifications, a development prototype board, and simple analysis of feasibility such as power consumption. I would have to write the software for developing and testing with the corresponding development board unit. This specification could apparently reproduce all the functions of my original idea; of having a display (using the then 'exotic' e-paper technology), a speaker, a paper-thin battery and a contact interface for programmability, all of which would be embedded within the form factor of an ordinary credit card. The feasibility study however did not include bespoke programming for testing various functions. Fortunately, Heng who was a brilliant electronics engineer and programmer with embedded system knowledge overcame the technical hurdles. Subsequently, he developed the PC interface without incurring further cost to the company.

I was surprised the technology of the Smart card alarm was simply one of the leading product provided by SmartDisplayer, the Taiwanese company whose honest and rather despairing executive I first met in the Cartes show the previous year. With the brilliance of hindsight and now with a bit more experience of the industry, it appears that with sufficient research and networking I could well have done much of the feasibility study myself! There was also another realisation that Taiwan was no longer just a low-tech manufacturing centre, but a growing powerhouse of innovation in new technology, rivalling the best in the world.

On the health front, 2005 was unexceptional except for having been re-referred to see a couple of new specialists. I started regular check-ups to see a gastroenterologist to monitor my liver, as tests had shown that it had sustained raised level of a particular enzyme, the name of which I cannot recall. This result alone did not necessarily indicate a serious liver condition, although it did mean the liver was not entirely normal. A liver biopsy was required to be certain. Since I felt fine, I did not see any drastic need to do anything about my Hep-C status. I was probably deliberately blanking it out from my consciousness because the fact is many H had already fallen seriously ill and some had died from it. Of course, I wished to rid myself of the virus as it was a stigma and acted as a mental brake to establishing a relationship. Powerful drugs had began to emerge which could potentially cure the disease. However, the treatment and its side-effects were infamous and notorious. The probability of failure and relapse was also high. In view of these odds, I decided against trying out the available treatment. As for my other H-induced problems, it was just another year of bearing it. I was put on the book of an orthopaedic knee consultant Mr Archer. The time to consider a knee replacement was not far off. In fact, when the X-rays of my bad knee were examined by Mr Archer, I heard him saying to his assistant that it was the worst knee joint he had ever seen! There was no cartilage left in the knee, and even the unprotected bones had been worn down. I wasn't really surprised as I was offered a knee replacement more than 10 years earlier. It was a miracle that I could still walk and not be in constant pain. Compared to an able-bodied person, my mobility had obviously been severely curtailed, but due to my stubbornness, I had accepted this limitation as part of my normal living. Subconsciously, I had conditioned myself to a restricted walking range, an abnormally slow pace and periodic bouts of immobility due to H. Other than that, I did not really suffer severe pain in the joint except when it was traumatised. I opted to defer the decision to operate as I was quite afraid of undergoing such a big operation. The knee had, in fact, become less troublesome as physiotherapists Ann and Thuvia continued their efforts to improve its condition by a mixture of exercise, gait analysis, new orthotic fittings and acupressure. Of these four, exercise and posture training were the most important. I can also confirm that acupressure treatment did definitely incited firmer and tighter feelings in the knee as though it had improved the blood flow circulation.

However, the problems with my ankles persisted, which also frequently impacted on the knee as I altered my gait to ease the pain. The physios had got the situation under better control, but for every occasional bleeds, the affected joints kept on accumulating damage. A referral to see an ankle specialist was made.

Progress was made, but 2005 was no personal annus mirabilis; it was just a reasonable start of an unchartered journey. As for Britain, 2005 became the annus horribilis of terror. Britain's involvement in the Iraq war had further radicalised many British Muslim youths and initiated the birth of Britain's first home-grown Islamist terrorists. Fifty two people died in London on the 7$^{th}$ of July from bombs prepared in Leeds by religious and brain-washed fanatics who thought Islamic doctrines justified the murder and maiming of innocent commuters.

## Optimism and creativity

With the help of Heng, the world's first feasibility study of the Smart card alarm was completed at the beginning of 2006. However, funding for the next stage was still elusive, and my repeated attempts to get Finkelstein to commit to his promise appeared to fall on deaf ears. He was always either away from the office or out in meetings. Thep helped me relay my requests but Finkelstein never personally responded. He did not reject me either, but I was beginning to doubt his words. I knew his company was developing a high security bank card which could display and generate a One-Time Password (OTP) security code. Such a card would reduce 'phishing' by fraudsters since a transaction could not be completed with just the knowledge of the card's PIN or password. The owner's card must also be present to generate the onscreen OTP. In the security world jargon, it is known as 'two factor authentication' which was increasingly sought after by banks and other financial institutions. There exists a huge market should all banks adapt this kind of card. Finkelstein's company had recently succeeded in raising a seven figure investment and was ready to expand. His project was obviously far more advanced than mine, but I was getting increasingly exasperated by his lack of action. The months were passing much more quickly than I wished. Without further investment, the development of my card would grind to a halt. I did not hang about though; I continued my investment quests and networked with potential clients and academics who worked in the Assistive Technology (or AT) sector. The latter brought words of support, but still the UK investment community I approached just did not get the opportunity or wanted to invest in a pre-revenue project. It was the classic chicken and egg conundrum again. However, I still had optimism and strong motivation that fuelled my entrepreneurial journey – Heng's parents, who owned a small and successful electronics company in Shandong, China, liked the business proposal of offering a portable programmable reminder for appointments and other applications such as medication reminders. China's rapidly increasing life expectancy opened a significant market opportunity for such products. The company did not have the expertise or facilities to make the appointment reminder in the smart card form factor, but they could make one that was the size of a match box using conventional electronic components. I accepted this offer wholeheartedly as at least we could soon have a product with the same functionality as the card (except without its compact form factor), and the opportunity to develop the software that controlled the alarm protocol and the user interface. The stress on my personal and company's immediate financial situation was relieved for at least a while more as I managed to borrow £40,000 by re-mortgaging my flat. This cash flow injection would have been impossible in the post-2008 financial crash as it was a self-certified loan paying interest only for the next 25 years.

Another development which kept up my motivation was that I had managed to make some fairly crude but fully functional 'iiCaps'. With these working prototypes, I was able to start raising interest for this invention as well. Once again I sought out another milliner who could make me some professional looking iiCaps, and perhaps also help me market them. I chased up various leads; from a Manchester hat manufacturer to a maxillo-facial surgeon, and even another investment opportunity on television; this time via a brand new Sky program called "The Big Idea". It was more like X-Factor than Dragons' Den. All one had to do was to turn up in the large convention room in Manchester, talk informally about my cap invention to the jolly and charming interviewer Di Stewart and the producer. Then I waited my turn to give a short pitch to the three judges: Craig Johnston (ex-footballer and inventor of the Adidas Predator football boot), Karan Bilmoria (inventor of Cobra beer) and Ruth Badger (2006 runner up in The Apprentice TV program). In return, they tried to knock the invention or the business just like Dragons' Den. There was a £100,000 prize money for the winner in the finals. I did not get past the first round which was disappointing as some of the inventions that got through were either not new or were just totally crazy. Despite not winning, I hoped public exposure may bring some unexpected benefits. As usual, there was a couple of months wait between filming and broadcast. Eventually, I got a couple of minutes of publicity out of it. Unfortunately, compared to Dragons' Den, this was a very lightweight programme which was shown to a fairly limited audience on the minority Sky 3 channel. Hence, other than a little bit of fun and encouragement from another TV appearance, I did not gain any meaningful response.

Not being overwhelmed by work may have been a disappointment, but it did have one big advantage – there was more time to think and be creative. Several years earlier, I had started to think about tackling a really private personal problem and fear that I had developed since Cambridge when I started my prophylactic treatment which requires inserting a needle into one's own vein. Having a rather overactive imagination, this process brings out a lot of anxiety that includes a fear of over-puncturing the vein, a fear of missing the vein, and of course the expectation of inflicting pain. When I was younger, I would, despite the pain and discomfort, innocently and fearlessly observe the injection procedure by doctors or nurses. Even when they missed and struggled to get the needle in properly, I was not afraid to look on as they poked and sometimes twisted and turned the needle in an attempt to find the vein. Now, I would just look away or even close my eyes. But when one is giving self-treatment, looking away is not an option. One not only has to consciously observe the process, one has to do it, creating an additional psychological barrier to self-treatment. With this in mind, I started thinking about creating an injection aid which minimised the anxiety of the injection process. An aid which allows one to inject into the vein without fear of over-puncture would certainly minimise some anxiety. It would be better still if I could invent an aid which could in some way make the whole injection process so certain and safe that one could do it blindfolded. The idea being that if one can perform the injection with more confidence, the anxiety aspect of the process would be reduced. After so many years of worrying over the process, I finally came up with a concept of making a special guide which allows a slightly modified needle (or cannula) to follow a conventional injection trajectory, but prevents it from slipping or accidentally over-puncturing the vein. Without going into the details of the design there was much to do to overcome several practical issues. At the time, I was not entirely confident that these practical issues could be overcome, but I felt this seed idea was probably on the right track which could lead to a new Intravenous Injection (IVI) aid invention.

## Progress is about Inclusive Innovations

With encouragement from the previous two developments relating to the iiCap and the IVI-aid, I felt like a genuine inventor of real potential! It was necessary then to create a new company which covered the above inventions and possibly other innovations which were not related to the Smart card alarm or its associated products. I thought for quite a bit about the name for this new company. One particular product caught my attention; Nintendo's latest console "Wii". The name had initially received universal derision; however, everyone was remembering it well, because it was short and sounded like a familiar word ('we') with positive connotations. Also, the letter 'i' was synonymous with cool technology products because of Apple and its ubiquitous products. With these inspirations and wanting my company's name to convey meaning and values I aspired to (which included Inclusivity, Inventiveness, Innovative, Imagination, Integrity), I named the company "Inclusive Innovations" and subsequently trademarked the logo "ii" so that when appropriate I could name my invention products preceded with the "ii" label.

The creation of the 'ii' brand was driven by necessity and mixed emotions; of great hope and vision as well as fear of failure. The iiCap, although 'simple' had the potential to become a popular consumer product. Similarly, if I could crack some of the design issues, the IVI-aid could become universally useful. There was the possibility and hope that through the success of iiCap, I could continue to develop new ii-products for the future, potentially creating a great company, like James Dyson had done. On the other hand, the fear of failure was also heavily hovering in the background all along. Both the Smart card alarm and the iiCap international ('PCT') pending patents were approaching their 18 months grace period deadline. The international search reports for both patents were not favourable, peppered with cited prior arts which apparently preceded my inventions. On my close analysis of these prior arts, I disagreed. But there was no way of getting an expert opinion without incurring high patent attorney fees. Time had been ticking much faster than I had wished for. The PCT deadlines were rapidly approaching which meant I must next progress the patents to the national stages i.e. submit each patent for examination in all the countries which I wished to apply. This was potentially hugely expensive as there is a cost of prosecution of each patent in each country. Having failed to raise new investment, my choice was either to abandon the patents or continue with a limited number of countries burning through my savings and loan. It was very hard to contemplate abandoning my patents after having invested so much work and money. Abandoning the patents would have felt like returning to ground zero, with all previous efforts and cost wasted. Continuing the patenting process would bring further financial strain but at least it would keep the hope alive that the jackpot was still possible. Psychologically, I felt compelled to continue the patent process. I opted to apply to the UK and USA for the smart card patent, and for the iiCap patent, I applied to UK, USA, and China. It was a calculated risk as the patents being granted were far from certain given the negative search and preliminary reports. Had I had sufficient investment I would have applied for patent protection in more countries. I managed to save several thousand pounds by doing the UK applications myself, and using a much cheaper online company to transfer the PCT patents to the corresponding foreign countries. It had still cost around £5000 in total. Hopefully, this investment would pay off in time, but I deeply wished the patents would not be examined too soon as I could not defend any objections without the help of a professional patent lawyer.

Failing to get new investment so far, my best chance of further developing the Smart card alarm still resided with Finkelstein's original promise. I returned to Paris for Cartes 2006 to continue my quest. It was an apprehensive time as many of the outer suburbs of Paris had been beset by violent rioting. My hotel was quite close to one of the riot areas but fortunately was not affected. There were also train and tube strikes which paralyzed much of the city, and prevented me attending one day of the show. On the days which I did attend, it was unusually quiet. If I was an exhibitor, I would have been very disappointed and angry that Paris in November seemed to be a particular bad month for strikes and riots, as these events occurred in the previous November too. I managed to meet Alan just once more. Having obtained some serious investment, his company had an impressive stand with Thep and several new members of staff present at the show. They had definitely moved on from the gimmicky light card to the serious new virgin market of high security authentication bank cards. As usual, he was difficult to track down. When I did find him, he was still vague but supportive in words. Some hope, but realistically his words had no bite. I was no longer trustful of his sincerity. Instead, my short term hope now lay with Heng and his parents to develop the ProActive Reminder. Frustratingly, that was also delayed until Q2 2007 as he took an extended 5 months wedding break in China over the Christmas and New Year period.

There were other tradeshows in the adjacent hall at the massive Villepinte exhibition centre. One of these was a fashion related show which I thought could be relevant to the iiCap invention. I might as well explore this show as well as there might be manufacturers who could help me develop the iiCap. There, I met many European and a number of Chinese firms that might be able to help, but due to lack of confirmed IP protection I was reluctant to explore further until I was better prepared. Then I saw the Hong Kong Trade Development and Council (HKTDC) stand, and started engaging with the staff. We spoke in Cantonese which gave me a sense of satisfaction that I was still capable of fluently conversing at the business level in the language of my birthplace. I was pleasantly surprised by the efficiency and helpfulness of the staff. Before then, despite being born in Hong Kong, I had not considered finding my solutions in Hong Kong or China as they seemed so far away and even a bit alien to me. The HKTDC also gave me catalogues of cap and other related manufacturers that could potentially fulfil the professional production of the iiCap. On my return trip from Paris, I chased up some of these HK companies, and was immediately impressed with the quickness of their response. I would email them in English, and get a respond by the next day. Even queries on technical and commercial matters were answered quickly and directly in contrast to many UK firms which tended to ignore my initial enquiries. Earlier in the year, and in the previous years, I had been in the process of partnering with two hat designers as well as a Manchester hat company, but all three of these opportunities did not materialise because the first two people couldn't take up the extra commitment and the latter showed no interest. I was sick and tired with this attitude; perhaps it was time to give up on these intransigent Brits and give the proactive Chinese manufacturers a try. Coincidentally, the North West Inventors club organised a seminar[140] about the pros and cons of manufacturing in HK and China. I felt reassured by the speaker's talk, particularly about working with HK based companies who still abided by the familiar and trusted British law system. Even with the cost of travelling and accommodation being taken into account, the cost of development and manufacturing was still a lot more affordable than doing it in the UK. In view of these benefits, the calling of Hong Kong was beckoning.

Away from business, my private life got a little more exciting as I took out a former student for a few dates to the wonderful Yorkshire countryside and seaside during the summer months. It had been a few years since I'd had the pleasure to be in the exclusive company of a female companion. Since returning to University in 1995, I'd not been particularly shy about approaching girls. My great friend Karim frequently reminded me that his second love of his life was through one of my initiatives in the then very popular 'Living Room' bar in York. However, for myself, because of my H and Hep-C positive conditions, and in addition to my conservative approach to dating, I was always excruciatingly slow in the usual dating steps. In today's world, it seems that you must kiss the girl on the second or third date at the very latest or the girl thinks you are weird, inexperienced, not manly or don't fancy her. Although kissing is very unlikely to pass on the Hep-C virus, there was no way that I could do it without first revealing my condition (what a great way to break up a potential relationship before even starting it!). I could defer telling the date about my H condition, but HCV was a bigger millstone. It would be much easier if the girl just fell in love with me so I could divulge my conditions more easily. So by my fourth dates, when I did not make the move, my hesitation had certainly killed off all sense of romance in spite of all my other wonderful qualities! However, this particular girl was unlikely to see me as more than a friend; as it turned out on that date, she confessed that she was already seeing someone else. Had I made the romantic move earlier, I may have swayed her mind, although I would never intentionally steal a girl from an established relationship. As it happened, I felt used, but we remained friends. Of course, I was also hurt, not just from the unrequited love, but by the persistent situation of my health status of living with H, living with a permanent physical disability which was increasingly restricting my freedom to explore, and now even more perilously the HCV condition inhibiting my freedom to romance. Without wanting to sound heroic, I did not feel any pity except a reasserted realisation that my honesty with my health conditions will always limit my chances of success with the opposite sex. Perhaps, I would have to consign myself to a life of a bachelor!

Coincidentally, at around the same time, my H doctor reminded me about getting treatment for HCV. He warned me that despite my apparent good health, the HC virus was known to be particularly dangerous after the age of 40, which I would be the following year. The latest treatment had also become considerably more effective with a higher success rate and was also less arduous to administer. One only needed to carry out one subcutaneous injection per week (instead of 3 times a week previously) and only one additional intravenous injection for blood monitoring. Previously, I had avoided the treatment because of the relatively low success rate and in my precious way, I could not stand the thought of more needles over a 6 or maybe even a 12 month period. With hindsight, I was really very naïve about the life threatening condition of being a carrier of the HCV virus – of the 4,670 UK haemophiliacs infected up to the 1990s, nearly 2000 have died by the time of the publication of the 'The Archer Inquiry'[141] in 2009. It was a no-brainer now; I wanted to get rid of the virus for good. I wanted to live without the stigma, without the fear of developing terminal liver disease, and yes, to kiss freely the next girl I date, in a normal way, without inhibition or anxiety. The process began with a consultation with a gastroenterologist at St James Hospital in Leeds in December 2006. He warned me about the side effects of the treatment which combines two powerful drugs; a weekly injection of pegylated interferon alpha and a thrice daily intake of Ribavirin. The treatment would be for a minimum of 24 weeks but may take up to 6 months depending on the response. The side-effects are extreme tiredness, hair loss, flu-like symptoms, and severe depression which could lead to suicidal thoughts! Interesting, I thought. He also warned me not to have unprotected sex during treatment and for at least six months after due to the treatment's toxic effect on embryonic development. I laughed aloud in my mind as that would be unlikely to happen.

Then, there followed a number of blood tests to determine my precise HC strain, as well as a precautionary HIV test. Even though I had been tested negative 10 years earlier, there was always a fear at the back of my mind that I could have been infected by subsequent batches of Factor 8 treatment[142]. The reminder of further nasty news was never far away as December 2006 happened to be the 25th anniversary of the first report of Aids, as well as news that vCJD was much more dangerous than previously thought for people exposed to contaminated blood products. There was no dramatic news about my HIV result; it was negative as before. I was also lucky; the HC virus in my blood was of genotype 2 strain which had the best response rate to the combination therapy. At last, the time to banish HCV had arrived. There was a waiting list of between 4 to 6 months for the notoriously unpleasant treatment. With luck, I would start my battle in April or May.

## Hong Kong, 10 years on

Having got some excellent leads from the HKTDC trade magazines, I prepared my first business trip to Hong Kong. The last time I was there was way back in 1995, in British administered HK. Although excited to be back at the place of my birth, I was quite wary about the Hong Kong of 2007. On my previous trips in 1992 and 1995, there were many wonderful characters of HK which made me so fond to be back, but I also recalled strongly its one big deficit; after being acclimatised to the general high standards of British civil manners (excepting the unruly drunks and occasional racists), to my shame HK people of the 90's era were not exactly the nicest people in the world. As I stepped off the plane, I was half expecting to be treated rudely by the customs people, get dismissed by strangers for enquiring and then get ripped off by the taxi drivers. As it turned out, the custom officers and the stranger who I enquired for direction were impeccably polite, and all the taxi drivers were friendly, with one driver even refusing to take any tips! Except for the smog due to the industrial pollution from mainland China, and despite the increase in population, the general environment in HK was exceptionally clean, even in bustling Mongkok. On my nostalgic trip to Waterfall Bay in Wah Fu estate, I was pleasantly surprised that the famous waterfall which gave its name was flowing with clear water and that there wasn't any rubbish strewn over its cliff or on the beach. It was more pristine than when I left it in 1978. Everyone everywhere I met on this trip were unexpectedly friendly and helpful.

My younger brother also came with me on this trip, as he was in between jobs after being made redundant by British Gas after many years of loyal service. This worked out quite well as it was always nice to have an easy going travel companion to share the experience, and he could help with my luggage should I have 'problems' with my joints or muscles. One of the weird things about travelling with my brother was my feeling of awkwardness when speaking to him in public. I was indecisive about which language to communicate to him with. In HK, it felt more proper to speak Cantonese to him, which he could understand, but mostly could not reply to properly. On the other hand, it also felt strange speaking English to him because in a self-conscious way I felt we were betraying our HK Chinese origin. Usually, I had no choice but to speak English with him because either I couldn't remember the Cantonese expressions or he wouldn't understand them anyway. Embarrassingly, when I did speak to him in English whilst in HK, my English fell to pieces with a strange hybrid of Cantonese and English grammar. This is an interesting contrast to when we are in England, when we only communicate in English.

Although there were friends and relatives who could provide accommodation, I preferred the independence and freedom of staying in a hotel near bustling Mongkok, where I could rest or carry out very personal IV treatment without explaining myself or needing to lock the room while in the host's apartment. I think all of my close relatives were aware of my H-status; however due to our separation and typical reticence of talking about genetic illnesses in Chinese culture I guess none of them really knew much about H or its treatment methods. So long as I didn't have any medical emergencies, that's exactly as I wished it to remain. Since my last trip, my great aunt and uncle (who sponsored my dad's first trip from China to HK) had passed away. A particularly wistful thing about being an immigrant's child in a country far away from your origin is the loss of much contact with relatives of all generations. It was wonderful to be able to unite with some of them as the passage of time would eventually sever these links forever. I particularly cherished the time with the gregarious Branda (my great aunt's daughter) who was always so humorous and generous. She had returned to full health after miraculously surviving a late stage breast cancer diagnosis a few years earlier[143].

Hong Kong no longer had much manufacturing. The cap manufacturers I visited had their factories north of the border in China and I had only visited their headquarters. The iiCap which I wanted to manufacture consists of two components; a cap which has some loopy fabric material (as in a hook and loop fastener) integrated on the underside of the peak of the cap, and a pair of detachable support systems which can be attached to the loop part of the cap. Disappointingly none of the cap manufacturers could accommodate making the simple support systems which I thought were relatively easy to make. I was particularly surprised that they did not know of the 'brim' reed material which I used for the support, which is used commonly in the millinery trade to give stiffness for couture hats. Apparently a cap manufacturer only knows about making caps. Even then only one of the manufacturers was flexible enough to make a cap with the slightly unusual structure I needed. Fortunately, I developed a rapport with the German marketing executive who eventually supported its development. The sample cost was extremely reasonable; just 50 US dollars per cap! One half of the iiCap creation was theoretically solved, but I would have to find another manufacturer who could create the support systems. I visited several more leads, but did not progress further as I did not have the finance to proceed or have the benefits of hindsight on the best design of the support systems.

I returned home refreshed and satisfied with the tentative progress I made, and the new experience I gained. I also reflected on how Hong Kong had changed in so many positive ways since my earlier trips. With the exception of air pollution, the city, its beaches and the environment all seemed to have improved dramatically. Culturally, it had grown beyond being famous for making money and movies; there was an unrivalled variety of both Western and Eastern offerings in entertainment, cultural events and local cuisine to enjoy. Intellectually and academically, three of Hong Kong's Universities are ranked in the world's top 100 Universities. In business, Hong Kong was already famous for free trade and its financial services, and becoming increasingly famous in the areas of the creative industries. As a hub for high tech and the creative industries, Hong Kong is perhaps the best place to take advantage of its proximity to the great centres of manufacturing in China. But most impressive of all was the civic changes; none of the strangers or public servants I met were rude; on the contrary they were helpful and chatty. Fortunately I didn't need any medical attention during my stay, but from what I had heard, this general improvement of Hong Kongers' attitude was across the board. Now, I had no hesitation to proclaim "I love Hong Kong".

Why and how had these positive changes came about? One ironic factor which led to the improvement in the environment was the effect of the bird flu pandemics in 1996/1997 and the SARS outbreak in 2002/2003. These disasters ironically led to much higher standards in health and hygiene measures. These outbreaks also caused an economic downturn, so people reacted more kindly to attract business. Government campaigns on better manners and respect for the environment during these uncertain times may have been another contributing factor. My own theory is better education and improving social development is the key to a more civilised society. My perception is Hong Kong's progress in civic manners is similar to advance in British society where casual racism is far less tolerated than just a few decades earlier.

The other great contributing factor is about politics and development. As 2007 marks the 10$^{th}$ anniversary of the return of Hong Kong's sovereignty status back to China, many people and commentators all over the world were interested in how HK has changed since the Chinese takeover. In my view, there was no doubt that HK has become a far better place in the way I have described above. The HK people must always take the first credit through their hard work, entrepreneurialism and pragmatic governance, with China and UK deserving credit for the following: China for its adherence to the 'One country, and Two systems' rule[144], for being the stalwart against the turbulence of the 1997 Asian financial crisis, which enabled HK to successfully repel the greedy short term speculators, and most importantly for being the key engine of growth for HK which, like the mainland, enabled it to enjoy higher growth and increasingly higher levels of living standards; which in due course led to a higher standard of civic engagement and demand for a better environment. Britain's credit is the legacy kind which has contributed to HK's success before 1997; they include beneficial factors such as the fully developed law system, nearly free NHS style healthcare system, freedom of the press, and the unique East-West characteristics which made HK so outstanding and attractive. Britain's other positive contribution to HK after 1997 was her non-political interference of her ex-colony, which combined with the pragmatic policy of China enabled Hong Kong to remain stable and successful.

# 9. THE LONG SALVATION

## Battles begin

A few weeks after I returned from Hong Kong, I had an inspired breakthrough to my injection aid which relied on a simple guide. The user can use this guide to enable the needle to be inserted more confidently into the injection site. But prior to my new idea, the guide had parts which obstructed the handling of the needle during the injection (or venepuncture) process. The inspiration came when, one day, as I was studying another needle from a different manufacturer, I noticed it had a safety feature comprised of a long sheath attached to the body of the needle that could be rotated to cover the needle after it had been used. Whilst playing with this mechanism, it daunted on me that the problem I had with my injection aid could be easily overcome simply by increasing the size of the handle which guided the needle.

Eureka! With a little more work, I could have a new invention! It was a wonderful moment and a boost to my morale as my other inventions were progressing slowly while the validity of the patents was hanging in the balance. The patents were worth nothing if they didn't pass the stringent patent office examination criteria. The examiners' reports for the iiCap and Smart card alarm patents duly arrived within a couple of months of each other (in February and April

2007 respectively). The news was bad; both applications had failed and I had four months to respond to the examiners. The news was not entirely unexpected. Despite my judgement that the prior arts were not relevant and that my claims were new and original, I just did not have the experience, skill and confidence to respond to the examiners. I had wanted to delay this moment for as long as possible until I had the resources to employ a professional patent lawyer to respond to any likely objections from the examiner. This is a common strategy for inventors, but my time had come and I'd not yet attained the necessary funding. I had to find investment soon but these negative reports had just made this quest even harder, if not impossible. I would have to respond myself, or employ an expensive patent lawyer gambling with my rapidly diminishing loan.

Meanwhile, the process of getting the Hep-C treatment was proceeding with a final pre-screening appointment with the specialist nurse. I almost couldn't make this appointment as I had just recovered from a particular bad episode of bleeds where I was bed-bound for over a week. During that period, I injected my precious veins almost daily. The injection sites had become bruised and its constant throbbing pain was a constant reminder of the need to preserve my veins. In my bedridden angst I decided I couldn't start the treatment as planned as I couldn't face additional jabs from the larger needle which was necessary for the weekly blood tests. Hence, as the smartly dressed specialist nurse attended to me, I told her apologetically that I wished to defer the treatment. Her reaction was instant and indignant. With a look of disgust, she threw my files onto the floor and proceeded to lecture me about wasting her time and about other patients being on the waiting list. This was a shockingly unpleasant reaction which I had never experienced in the NHS. She could have at least asked about the reason before treating me like a despicable time waster. It is true that many people do take the NHS for granted or even abuse it, but of all people, I was most sensitive about my indebtedness to the NHS and therefore had been most diligent in my relationship with it. I had enough guilt already and really didn't appreciate someone in her position putting me down like this. I felt angry but numbed. I wanted to lecture her back, to tell her to be more compassionate; instead I just left the room calmly without saying anything I may later regret.

I was already in a heightened mood that day. As whenever I had a major debilitating bleed which housebound me for a while, the ability to move freely or go outside again always led to an exaggerated sense of freedom and self-awareness. Did the nurse react the way she did because I am Chinese? That day, whilst waiting to be seen in the hospital waiting room, I felt particularly sensitive about myself and the thoughts of other patients about me because of another event. Two days earlier, a crazed Korean-American student had murdered 32 people at Virginia Tech University campus. Even though this tragedy had nothing to do with me or with my ethnicity, the constant broadcast of that despicable East-Asian killer heightened significantly my race relative to the other waiting patients. For some moments, I imagined for a self-aware person, it must be so much more uncomfortable to be a Muslim, or Black minority who had to endure frequent negative reporting from the media.

As the sore of my veins dissipated, the desire to exterminate the Hep-C virus returned. An advantage of having the 'H' condition is the specialist care teams that are available in the major health centres. Instead of going back to the angry nurse at the gastroenterology department, my treatment would be under the care of H specialist nurse Angela. With her natural black hair and pretty look, she reminded me of Sister Shaw who had looked after me so well in my childhood. As a specialist nurse treating H patients for many years, she was brilliant with the needles. I felt totally at ease with her.

All I had to do was to turn up weekly at the clinics (for possibly up to 6 months), and she would administer the subcutaneous injection into my tummy or thigh, and carry out the IV for the blood test required to check if the powerful drug had adversely affected my blood count. My first treatment would start on the 31$^{st}$ May 2007. With luck, I could be cleared of the Hep-C virus by autumn. Without luck, the treatment could go on through Christmas, and I could become an emaciated balding manic depressive wishing I had not bothered.

Running up to this first treatment date, I was busily working through the patents to see if I could respond to the examiners through my own analysis. I had to do as much as possible as there was a likelihood that I might not be able to work during the treatment. Despite knowing my inventions intimately, it was probably impossible to objectively and effectively counter the judgements of the examiner without the experience and skill of an independent expert patent lawyer. There was a real sense that I could lose my patents, and wasting all the time and money invested. There was one lifeline I could try. A few months earlier, Suzy, whom I had helped a few years earlier in her business development, had sent me the CV of a patent lawyer called James Love in Harrogate who had recently set up his own practice. They were acquaintances who had met through a networking meeting, so there was no personal introduction. I didn't bother to contact him earlier because his CV put me off. He was an award winning lawyer who had previously held two top posts as Head of IP at two leading IP firms in Leeds. As I couldn't afford to pay much, it seemed unlikely that someone in his position would bother to help me in my precarious situation. But I had nothing to lose. I called him and an appointment was set up in April, where I met him in his modest office. My immediate impression of James was superficial. I noted his kind face and light blonde ginger hair. His voice was slightly high pitched but soft and assured, definitely posh but with a hint of northernness. His body language was humble and almost diffident. There was no sense of cockiness or pretentious confidence that I had come to expect from my previous experience of some young lawyers who I had met in the past. I needed to impress him as I would be asking him to consider my case on a contingency basis or some investment type arrangement. Effectively he would become an investor should he decide to help me out. I showed him all of my inventions, vision, and ambitions; I was also honest about the uncertain IP situation which I now found myself in. He listened carefully and gave me plenty of time to express myself. Obviously, he needed some time to consider my patents' potential and my proposal. There was a nervous three week gap to our next meeting as he went away on business to an international trademarks show in San Francisco.

James had been impressed with my commitment and what I had done almost singlehandedly. Despite the apparent difficult objections raised by the examiners against my patents, he calmly gave me his balanced assessment. There was hope, but no guarantee of success either. Yet, with great generosity of spirit, he agreed to help me out in preparing the response to the examiners, as well as other IP matters. This was the best news I could have wished for. I couldn't believe my luck that I had got an award-winning IP lawyer to help me out, at this most critical of times in my invention journey. There was a palpable sense of relief. Otherwise, the stress of trying to save my patents would have been overwhelming, particularly because I was about to start the mentally and physically tough combination therapy to eliminate the HCV virus.

The first course of the treatment was on Thursday on the 31$^{st}$ May 2007. Angela gave me the Alpha-interferon jab, followed by two tablets of Ribavirin. There was much anticipation and anxiety about the drug's infamous ill side-effects. I recall vividly sitting by myself in a small Chinese eatery in the working class Merrion shopping Centre in Leeds as I waited for the side-effects to hit. Nothing happened as I tucked into my delicious spicy stir fried noodles, washed down with a can of cooled coke; my taste buds heightened by the highs of finally embarking on a life-changing treatment. Then, suddenly, clear fluid dripped from my nose. I wiped it away quickly with a tissue paper. Then my nose went stuffy like a cold had hit me. I wondered if this was the start of flu symptoms commonly associated with the side-effects of the medication.

Fortunately the stuffy nose was just transient. With the exception of sharp tingling pulsations on the back of my head and the weird sensation that time seemed to be passing more slowly than usual, I felt fine for the rest of the week. Funnily, I was a bit disappointed for not experiencing more severe side-effects partly due to my unscientific thinking that the strength of the side-effect equated to the body reacting or fighting the virus more strongly.

In week two, a small argument over the fate of some old furniture with my dad led to me to being extremely angry and intensely resentful of him. What a stubborn old git. Why did he not understand or appreciate my condition? All the anger and resentful feelings just built up volcanically as I drove from my parents' house in Manchester back to York. It was a strange feeling of high emotions apparently over something trivial. This state of intense negative mood, resentment and dark thoughts lasted the entire weekend. This must have been the side-effects which I had been warned about. The drugs had apparently magnified my mood and prolonged the bad feelings. It was a good thing that I wasn't a depressive person to start off with, else I may well have harmed myself. The rest of the first four weeks of treatment passed by uneventfully. This was a positive milestone as it meant I was unlikely to drop out of the treatment course due to adverse side-effects. Such is the power of the drugs, that on week five some blood was taken from me to check if there was still HCV in my system. Although I had started feeling a new symptom of extreme tiredness, I was very excited and anxious of discovering the results in a week's time.

After what seemed like a time-retarded existence, the day of reckoning finally arrived. The effect of the drugs were still somehow stretching and terribly prolonging the passing of time. It was not a good day anyway as I had stayed up very late the night before, discussing the fate of my cap samples I'd ordered from the Hong Kong company a few months earlier. They had apparently gone missing in the post and the whole process of remanufacturing them had to start again. As I awaited my appointment with the doctor who would tell me the results of the HCV test, I prepared myself for the bad news. If the virus had not already been cleared, there would be another five more months of treatment to endure. If the previous week was a guide to how time passes in a week, then five months of further treatment would probably feel like 3 years of incarceration. It was a new junior doctor who saw me that day. We had never met before, so he spent rather a long time mulling over my records, whereas I just wanted to find out the test results. Eventually he announced the result: the result was negative, there was no longer any indication of the Hepatitis C virus, and my liver function had returned to normal. I paused for a few seconds, trying to take in the news. Then, I gave out a burst of great joy. I wanted almost to laugh out aloud, to hug someone that appreciated what I had been through. But I didn't know anyone there, so my celebration was subdued. However, when I returned home, I almost felt hysterically happy as the full implication of my new cleansed status began to sink in. I became emotional thinking of the years of restrictions that I placed on myself. The anxiety, the stigma, the missed opportunities and real consequences of living with a potentially deadly virus had finally been lifted out of my subconscious. It should have been one of the happiest days of my life, but I was too drained from the treatment to celebrate it.

The positive news meant I would only need to carry on the treatments for just three months more instead of the full further five months had I not responded. These additional months of treatments should eliminate all remnants of the virus. The next three months were not plain sailing; tiredness and lethargy were constant companions. Additionally, parts of my hand had developed blistering and dermatitis; part of my legs and waist had developed eczema-like rashes; and most worryingly of all, in a vanity way, chunks of my hair had started to fall out in the final months of treatment. Also there were periodic bouts of depressive thoughts and at times intense anger. According to my diary, there were even moments of rage as I dealt with the delays and frustrations in my business developments. To my shame, I even resented my mum when she needed my language support during her extremely painful gallstone illness. These feelings were exasperated through the increasingly troublesome ankle pains I was experiencing which subsequently led to a bout of intense shoulder and back pain as a result of my compromised gait.

Ironically, these few months of roller-coaster emotions were also my most creative as I solved the final stumbling block to the design of the IVI-aid when I successfully formulated a mathematical model of the injection path which enabled me to design an optimally shaped guide. Also, I wrote the opening paragraphs of this book during this time, which boosted my confidence to write. Furthermore, I even managed to work under pressure as I prepared as much relevant information as I could to aid James to defend my patents. Fortunately, despite my mood swings, I had remained fully professional and diligent throughout. When James completed his analysis, the reply for each of the examined patents was just a few pages long. There was no fancy jargon or complicated arguments; the examiners' objections were just addressed directly and concisely. On first inspection, I thought to myself "Is that it!?" The reply looked too simple and straightforward, because the real skill was the ability to explain the key point of an argument based on the existing description of the invention. Of course, on closer analysis, I realised that James had expressed most of the points I wanted to raise, plus some more which I had not realised. He assured me those points he missed out were superfluous or could be used in subsequent arguments should the examiner not be satisfied with the reply. The fate of both of my patents now lay with the judgement of the examiners.

With regards to the rest of my business developments, Finkelstein's company InCardtech had expanded significantly through an IPO which saw his company taking over the French company Audiosmartcard, along with some of the joint IP we developed. I complied readily with their request for a 'transfer' agreement to demonstrate my efficiency and flexibility, hoping that they would respond in kind. In the preceding Cartes exhibition in Paris, they had a very impressive stand in the show with Thep, Nick and several other assistants there. Their authentication card was finally ready and they had been marketing it extensively. They were busy engaging with potential clients and I felt I was a bit of a nuisance despite managing to get some more reassuring words from Alan that my project was still in hand, though I did not trust him anymore. I returned home and tried chasing up Alan again for a decisive agreement. Some progress was made with his lawyer as we started to negotiate terms, but as the process went on, it became painfully clear there was no hope. I gave up on him. How disappointing that people who say they would help don't stick to their word. With hindsight, I wished I'd held out on the transfer of IP agreement until I got what I wanted, rather than being all nice and cooperative.

I was also frustrated by the lack of communication between Heng and me as he travelled back to China for another long break. With the sample cap finally produced, and further progress made in the support system's design, I was increasingly drawn back to developing the iiCap. With my 40$^{th}$ birthday fast approaching, I worked feverishly to complete the patent application for the IV-injection aid. I did not care how commercial this invention would become, but I was certain that it would definitely be useful for some people, particularly H's. It was a very satisfactory invention to have developed because it had sprung from my most personal of experiences and made use of my physics background too! With help from James, my latest invention patent was submitted on the eve of my birthday. A well-deserved present to myself, I thought. By coincidence, my good friend Karim was in York. In our favourite bar, Karim generously bought a bottle of Champagne to celebrate the beginning of my fifth decade.

Most people reflect about their life on this milestone date; about relationships, children, and career. I can't say I gave any of the above much thought because I was still at ground zero on all three fronts. My inventions were my substitute babies and I hoped they would grow up fast and be successful. There was no doubt that my medical conditions played a large part in my uncertain state. Despite all the setbacks and delays, I remained a pretty cheerful chap; living with H had conditioned me to be patient and optimistic.

## New blood, new knee, new life

Although frustrated business wise, on the health front there was one great success with the elimination of the Hep C virus to perk me up. It was like my blood had been reformatted, giving me a fresh and brand new outlook on life. Physiologically, I honestly could not say it made much difference then, although it would be certain that had I not got rid of the virus, I probably would have become seriously sick later on. The biggest benefit was psychological, I felt cleansed with one less burden of anxiety and guilt. Potentially, the most dangerous roadblock of my life had finally been vanquished!

There was one more troubling roadblock to sort out though. The severity of the condition of my right knee had already indirectly inflicted much discomfort to other parts of my body. My ankles had been especially enduring of the abnormal strain placed on them in compensating for the deformity and weakness of the knee. I should have had knee replacement surgery ten or more years ago, but I was too much of a coward for surgery and had put it off for as long as possible. The decision of Mr Archer, the knee surgeon to soon retire finally persuaded me to take on my next big medical decision. Mr Archer was obviously a very experienced knee surgeon, with an excellent record and experience of performing knee replacement surgery on H-patients. Better get this operation done with him than with an unknown quantity! I was placed on the waiting list. Meanwhile, after a ridiculous four years wait, I finally received an appointment with the ankle specialist Mr Harris. From the X-ray scans, arthritis caused by H was evident in both of my ankles. However, there was not much he could really do until the knee is sorted out first. My operation was scheduled on Valentine's Day, 2008.

The day prior to the operation, I kept myself busy by doing DIY in the loft of the house before driving to my parents' home in Manchester for the evening before the operation which was scheduled for the morning at Chapel Allerton Hospital, Leeds. Unlike in the old days, there was no overnight pre-stay to acclimatise the patient. Nowadays the principle is that the fewer days the patient is in hospital for, the better – there is less chance of infections while patients generally remain less stressed at home than in hospital. Furthermore, it's more efficient and cheaper for the NHS. My younger brother drove the car to the hospital, and we arrived pre-dawn around 6.30am. Upon arrival I was told to have a clinical shower and wait in a rest room. I told my dad and brother to go home rather than sitting around for hours as I feel more relax being on my own. The situation might have been different had I got a partner; it was Valentine's Day after all. Specialist H-nurse Angela arrived and we chatted casually which eased my nerves as she effortlessly inserted a cannula in my left hand. A much bigger dose of Factor 8 than usual was then injected into my vein, in preparation for my full knee replacement – a substantial operation which I tried hard not to think about. There was the butterfly in the stomach feeling all over me now. To be frank, I rather liked the feeling as it's usually associated with travels and adventures. In a way, I saw this operation in a similar light because nothing else mattered except this operation and my recovery; all the worries and frustrations with my business and the patent deadlines were distant secondary matters.

Mr Archer, the surgeon came in with the usual pre-op prep. I smelt tobacco; he must have had a cigarette earlier[145], which in a strange way reassured me a little as it reminded me of relaxing time in pubs with friends. Although he must have already known the answers from my pre-op forms, he asked me a series of questions like have I got any infectious blood conditions such as HIV or HCV? I was glad to tell him "No" in both cases, emphasising that I was recently cleared of the latter. I detected a sigh of relief from his side even though it would not have prevented the operation. Another question he asked was quite strange I thought: "Which leg are you having the knee replacement done"? A momentary mischievousness filled my thought of such a silly question – jokingly, I thought briefly of telling him "It's my left knee. Haha"

Well, I didn't; it would not be so funny to wake up with my good knee replaced instead! The time finally came for the operating theatre. A male nurse came in. I was expecting a trolley bed that would transport me there, but instead the nurse and I casually walked to the theatre room, which was just around the corner. This was going to be my last walk with my knackered knee. More butterflies were set loose as I entered the room. Another surprise… I had expected it to be a brightly lit operating theatre, clinically white with doctors and nurses all around. Instead, it was like a store room with shelves all around, except there was a trolley bed in the middle. To my left, in front was Mr Archer sitting with his back to the bed, on a standard office chair, casually sipping a cup of tea. It felt rather surreal.

Ah, I see, I was actually just in the anaesthetic room. I think I only had a glimpse of the operating room, in a dream perhaps.

I woke up around mid-day, with an oxygen mask over my face and left hand on morphine and antibiotic drip, whilst the forefinger on my right hand was clamped with an oxygen and pulse monitor. My right leg was almost completely covered with heavy bandaging, with the exception of a drainage tube carrying blood away from the operated knee into a bag. The first familiar person I saw was the reassuring and smiling face of Angela; my immediate thought was 'I have finally done it', but the first thing I actually mumbled was "I am still alive!" Modern medical procedure is of course very safe in general, but there is always a chance of unintended complications or that something can go wrong. My overactive imagination and anxiety had definitely delayed this inevitable operation, but finally I had done it. Hopefully, all would be well and my recovery would go smoothly.

There was another H called Bob who also had a knee replacement done on the same day. His bed was next to mine so we had much opportunity to chat about our lives. He was in his early sixties, so he grew up in the days before effective anti-H treatment became readily available. I could see this through his battered body. This was his third knee replacement. Besides that, he'd had two hip replacements and operations on both elbows. He needed a Zimmer frame to get about as all his limbs were severely damaged by the cruelty of H. He had other serious health conditions too which I would rather not divulge. His brother who also had H had died. I didn't asked how, but it was obvious that it was due to H. Bob was a tough guy, a survivor who must have gone through hell in his younger days before modern treatments became widely available. Despite his terrible disadvantage in life, he had a supportive family and a really uplifting character. Therefore I felt happy for him as a supportive family is so important for anyone who has to endure such a difficult life, while a positive outlook on life is so necessary for contentment despite the frequent punishment caused by H. I don't know if this uplifting quality is a peculiar trait of H, because it seemed to me that almost every H that I have met have shown this resilient and happy go lucky attitude.

In the old days, a full knee replacement even for normal patients without H would have meant several weeks, or even months in hospital. Also, absolute long rest was usually prescribed for fear of harming the new knee replacement. But, today's protocol is almost exactly the opposite! On the very next day, the patient, guided by a physiotherapist is expected to be out of bed and made to bear full weight on the joint. Bob did just that, whilst I was excused from it because the swelling on my knee was still exceptionally bad. The next day, I did briefly touch down on both feet. It was an incredible feeling, putting the full weight of my body on a severely swollen knee, yet without feeling any additional pain as I would normally expect before the operation! Slowly, I also started to bend my knee a little, but the swelling prevented much movement. In order to have a good functional knee, it needed to be able to flex between a straight position to an angle of at least 100 degrees. Before the operation, my knee, despite its severe damage had a remarkable range of 130 degrees. The second and the third day after the operation were also most intensely excruciating: the nerve block that was applied during the operation had begun to wear off. The resulting pain was something which even I had not felt before – there were simultaneous and incessant waves of electric shock-like pain from the foot, mixed with an intensely deep numbing and dull throbbing pain in the knee. In times like this, I couldn't but help myself by quietly moaning "God… oh God…aah God it's bloody hurting…" I was not religious, and since my enlightenment experience almost exactly 15 years earlier nor did I care if 'God' exists or not, but I have to admit that during such times, there was something powerful and soothing, and perhaps even yearning when that word was called out. There were also other anxious thoughts; for some reason, I kept imagining the suffering of injured soldiers and civilians hurt in war and disaster zones, without analgesic and with limited medical help[146]. Ironically, there was a rare and significant earth tremor a few days later while I was still recovering in the hospital.

Meanwhile, Bob's recovery went remarkably smoothly. Despite his age and significantly worse overall condition, his knee's swelling subsided rapidly, so by the fourth or fifth day his knee's range of movement was almost back to his previous normal. Meanwhile, my leg's swelling was still so severe that I had an ultrasound to check for deep vein thrombosis; a common complication after KRT. Fortunately, none was detected; instead it appeared that my body was burning up Factor 8 at a much faster rate than expected. This, together with my particularly reactive body metabolism had led to sustained inflammation in the operated joint. One full week after the operation, Bob was well enough to go home, in the same time frame as a normal person without complications. I was very happy for him but grew increasingly anxious about my own situation. The knee and my whole leg was still very swollen, red and hot to touch. Every morning when I woke up, the swelling was subdued, but as the day time pass by, and especially after some mild physiotherapy, the swelling would return with a vengeance. This high and low cycle, which was also accompanied by nightly sweating, just kept repeating itself, day after day. Hence, there was hardly any progress in the knee's range of movement. It just would not bend beyond 50 degrees. A knee that restrictive meant I wouldn't be able to drive a normal car, or sit down properly without the vulnerable leg sticking out, or be able to go up or down stairs normally. Most worryingly, I was warned that if the swelling didn't subside soon, H patients in particular had a tendency to develop scar tissues within the joint. If this happened, then the knee would become extremely tight and rigid. I had to get my knee bending fast or it would be stuck there. But I could not start bending the knee more until the swelling died down.

Despite everything, my time at the hospital was otherwise quite 'relaxing'; I forgot about work entirely and didn't have to care about chores and other distractions like news, TV or the internet. I had finally completed the reading of "The Fabric of the Cosmos" by theoretical physicist Brian Greene. The book covers an amazing amount of physics in a densely written style which most lay people could actually follow – if they tried. It was an incredible tour de force of just about all the important concepts in physics. Reading it in my hospital bed, had at times taken me to the edge of the Universe, into the unfathomable subtleties of the atomic quantum world, and to the beginning of time. Beside one's own imagination, other patients and I got to know each other, and I had the occasional chitchat with nurses and physios when they were less busy. Unfortunately, there was no chitchat with doctors though because, as ever, I hardly saw them during my in-patient stay – I would say the total consultation time with Mr Archer was probably no more than a few minutes. I suppose he had done his main (and excellent) job in the operating room, but it would be psychologically more assuring if every patient had an extra few minutes talk with their doctors. Hospital can be a very good place to socialise and make observations of human relationships, especially for seeing the caring and cosy side of them. In particular, I think the ward environment of hospital is possibly the most common place where ethnic minorities, immigrants and indigenous locals get to learn and observe each other's family relationships. Contrary to the brusque reputation of the Yorkshireman, my outstanding observation in this hospital stay was the (almost excessive) overt display of polite manners of the patients. Of all of my hospital stays across the country, I have never heard so many "please" and "thank you" in my life.

I was finally allowed to go home after exactly two weeks at the hospital. Being at home was definitely much more relaxing, but the frustrating swelling cycle of the knee remained unabated, which meant I hardly made any progress in being able to bend my knee more. It was absolutely crucial to persist in the physio sessions to strengthen the leg and to gain a greater bending range. It is a sad fact that a proportion of patients from all backgrounds (not just those with H) think that surgery alone will magically cure them. The reality is post-operative therapy is just as important, which if neglected can lead to disappointment and even failure of the procedure. Therefore, I pushed myself very hard every day to try to regain the range I had before.

But after a further two weeks at home, the swelling cycle persisted and my knee still could not bend much more than at the beginning. I began to think that scar tissue must have started to form and that I would never achieve the range required for ordinary function. The situation was very frustrating and worrisome. There was an air of resignation that this operation had failed. I thought about selling my car and coming to terms with my new state of being. It would have been a big disappointment for such an outcome.

Then on my forth week post-op, I did something which I used to do commonly during the daytime. I put on a mildly compressive 'tubigrip' bandage over the knee during sleep to try to suppress the nightly swelling. To my surprise, the swelling was noticeable less in the morning. Over days, the swelling cycle was finally broken. I felt elated, and astonished. Subsequently, with the help of diligent exercise and physiotherapy, my knee eventually regained a maximum bending angles of 100 degrees. Not perfect, but I could do pretty much most normal activities, including even cycling again four months later.

It was a long recovery, but it was worth it. Life is so beautiful and free when one can just walk normally without discomfort and constraint. I could start to learn to walk with a normal gait again, just in time to prevent developing further back and neck problems. York is a small and beautiful city with many hidden treasures, but I hadn't felt compelled to explore these wonderful places until now, without the shackle and chains of a too often debilitating and painful disability. With the new knee was a new energised me; if I wanted to stand a bit more, I could. If I wanted to walk a bit more, I could. If I see something beautiful or unusual, I would explore it without my desire or curiosity extinguished by pain and immobility. To love and being loved is no longer such a high barrier. Despite the still inescapable reticence about H, I even found a beautiful and good hearted girlfriend not long after[147]. Life is once again simple, and wonderful!

Except... all was not perfect. With the elimination of my dodgy knee came the realisation that my ankles and feet were now my nemeses. Years of compromised gait and unnatural strain on them had already caused me almost as much grief in recent years, except the problems were masked by the problems of my pre-op knee. Frequently, the ankles would even lock up literally and suddenly stopped me in mid-motion. Oh, bloody hell, when would I ever get a frigging break!

Ok, I don't want to bore the reader with more stories of my debilitating conditions, except to say that they were eventually resolved (mostly or at least brought under control) when the excellent H-specialist physiotherapists and chiropodist team at St James hospital Leeds combined their expertise to diagnose and mitigate my increasingly problem lower limbs.

Incredibly, a pair of new shoes with unusual curved sole and bespoke insoles did the job. The latter helped position my foot in a less stressed position while the curved sole reduced the dynamical stress on the ankle during walking. Brilliant!

Work wise, I continued my quest, setting up meetings with potential partners and going to conferences and exhibitions to network and find out about funding for innovative projects. I remember one event particularly well. It was the National Institute of Health for Health Research conference about a new i4i ("Invention for Innovation") grant held at the Queen Elizabeth II Hall in London. I was sitting in the audience listening to a talk about invention and IP protection when I received a text message from James, who a year earlier had helped me to file a response to the examiner concerning my iiCap invention. The text message was "Kin, your spectacle support system (iiCap) patent has been granted". The patent examiner had finally replied. I sat there, stunned. After almost four years of uncertainties, anxiety, financial and mental sacrifices, and fear of failures, I had been proven right! I had officially become a bona-fide inventor with my first granted patent. All was not in vain. I sat through the rest of the talk feeling jubilant and triumphant. Later, I rang James to thank him profusely. His apparently 'simple' reply was spot on – no faff, just clear concise and incisive answers to every query and objection from the examiners.

But soon after, that joy was tempered by the first examination reply from the Chinese patent examiner on the same invention. The patent application was rejected on similar grounds as the earlier decision of the UK examiner. James calmly advised to use the same arguments that he used in his previous response. Hence, I submitted this to my Chinese patent lawyer, hoping that they would be table to translate James's arguments as powerfully and as succinctly as the original English response. I wondered how long I would have to wait for this crucial reply. This is the big one as China is likely to be the largest market and source of manufacturing for the invention. Hence getting this patent was crucially important and potentially very valuable.

I didn't have to wait long. Less than three months after our response, the Chinese examiner replied. The Chinese patent lawyer didn't give me the decision directly in the email. They translated the opinion in a letter, and it read *"...we received a notification from the State Intellectual Property Office [SIPO] of the PRC in connection with the captioned application, which informs us of the following: Upon substantive examination, the Office found no cause for rejection of this application, and has decided to grant patent right"*.

Wow, Wow, Wow! I was absolutely delighted. Looks like my invention journey was definitely back on track. Also, by that time I had managed to find a HK manufacturer who could make the spectacle support parts of the invention. It wasn't perfect, but at least I now had a completely functional version of the iiCap for demonstration.

I hadn't given up on the Smart card alarm either; I prepared one more trip to the 2008 Cartes show in Paris. I went there to seek Philippe's advice (the former VP of Audiosmartcard who first helped me out) and to check on the latest developments in the field. Philippe had become the CEO of his own company in the innovative smart card development business. I trusted him as he had been consistently approachable and helpful during our interaction, whereas others had let me down or ignored me. I didn't see InCardtech again. Apparently Finkelstein was no longer the CEO and the company was burning cash at an unsustainable rate. With large investment came VCs who were ruthless in purging out founders who failed to perform. Later, I learned

InCardtech went into administration. Selling new technology is hard at the best of times, but with the onslaught of the 2008 financial market crash, failure was inevitable. I would just have to continue to find investors back in the UK. With the biggest economic crash in recent history, the prospect of finding investment became virtually impossible. In finance jargon, this was known as the valley of death, referring to the common fate of so many start-up companies which had to die because of the drought of funding. After several more failures to gain investment, I decided to focus on my iiCap invention instead.

## Dragons' Den Sport Relief

In 2009, the BBC's infamous business programme Dragons' Den had gone online and invited anyone to post a clip of their ideas which would then be voted on by online viewers. This was an excellent free way to get some market research and feedback of the iiCap. Hence, I prepared a video to upload onto the site. I had no intention of applying to go on the TV version of the programme again, as I knew how haphazard the whole process was; one could spend so much time and effort on the application process and still be omitted in the broadcast. But just as I was about to upload the video, I received an email from the Manchester Inventors Group alerting me to a special charity edition of Dragons' Den, focused purely on sport related products. It was going to involve celebrities who would act as advocates to the contestants' inventions. Suddenly, it felt really enticing to have a go with this opportunity as the iiCap was primarily aimed at sporting and leisure activities. I applied again, and was subjected to the same rigorous process; auditioning in Manchester, followed by weeks of interrogation and uncertainties. I worried I might be disqualified because one of the questions was whether I had demonstrated the iiCap on TV before. I answered honestly, playing down the significance of that rather forgettable Sky 3 programme.

Surprisingly, I got through again! Out of nearly 80 applicants, about 10 products were chosen, an Irishman with a bike which moves sideway like a crab, a couple of ladies with an exercise pillow, a football goal contraption, a Swiss guy with a satnav system for skiers, and several others with new sporting equipment. This time, the show was filmed in the legendary Pinewood studio in Buckinghamshire. Most of the contestants stayed overnight in a posh Heathrow hotel ready for the next day's shoot. I say posh because I recall the bathroom had shiny expensive marble floor tiles which caused me to slip after the shower. Fortunately I stopped the slide just in time to prevent a serious accident to my operated leg. A lucky escape indeed, as I thought why do so many hotels choose to install extremely slippery surface in the bathroom? In the early morning, we met other contestants waiting for our transport to Pinewood studio. There, I saw this big eccentric looking guy with a wide moustache who looked very familiar. I told him, "I've seen you before… yes that's right you were on that Sky 3 show 'The Big Idea', right?". The guy immediately put his finger in front of his month and said in a hush voice "Shhh... I haven't told the BBC about it". I told him "I was on the programme as well, but never got as far as you". He was the Irishman with a totally useless sideways bike. It looked as though the BBC wasn't really that diligent when it comes to checking the background of the contestants. I didn't mind at the time, but I did feel that he had cheated the system, depriving another more deserving contestant with a more serious product. Two familiar black people carriers picked us up – I was sure I had seen the same ones used in the BBC's other famous reality programme, The Apprentice. In the car I sat next to an impressive middle-aged businessman who spoke with a continental accent, probably German or Swiss. He carried with him an impressive cargo of the latest phones and other handheld devices which act as satnavs for skiers. Both his products and his professional manner were very impressive. As we talked about our businesses he showed me the HTC Kaiser, the latest high spec smartphone which coincidentally I had been thinking of getting. He let me play with the phone and I expressed my desire to get one soon. Then to my surprise, he said "Take it, you can have it"! He had a boxful of them with him. I thought he was joking and gave it back to him as we parted ways to prepare for filming.

I didn't have to wait long as the order of filming was me first. I wished it wasn't though because there was less time to settle down in the studio, to relax and snack on some delicious BBC food! More seriously, going first is probably least advantageous as the Dragons are less likely to invest, betting there might be better proposals to follow. Unlike the ordinary show, we (James also came along as my IP advocate) would first meet our celebrity supporter to prepare with him (or her) before the filming in front of the Dragons. A BBC employee prepared us for the meeting, she said "You may not recognise the celebrity, just go and greet each other and introduce yourself". When the door opened, a slightly portly young man with a kind smiling face walked in. He was wearing a girlie pink T-shirt and casual blue jeans. My mind desperately tried to figure out who he was as it would be a bit embarrassing not to be able to recognise a celebrity! I was also a bit disappointed that he wasn't a really famous star which everyone recognised instantly. Anyhow, he introduced himself before I had a chance to say something silly like "Sorry, I dunno who you are?" He was Jason Manford, a comedian. Instantly, I felt more relaxed with his layback style and familiar northern accent. In fact, he was from Salford/Manchester as I found out later which made me feel more akin with him. As a comedian, I had expected him to be cracking jokes all the time and filling the air with hilarious laughter, but no, he was rather matter-of-fact, composed and quite reserved! I suppose those documentaries about comedians' real life personalities being rather shy and introspective were true after all! Funnily enough, pairing a comedian with me was quite apt as I had imagined being a comedian myself when I was a younger. In fact, whenever I meet old friends from school, University and work they often remind me of my old antics and sayings which had them in stitches. Perhaps I was a natural clown as I hardly recall anything I said or did as being deliberately funny. What I find peculiar though, is how so many of the British comics look Chinese – Michael Mcintyre, Stewart Lee, Paddy McGuiness and even Jason Manford too with his round chubby face and slightly Chinese eyes.

So we prepared for meeting the Dragons as I explained to him how the iiCaps work. There was a minor disaster as a part of the prototype broke while Jason was playing with it. Fortunately I had some spare parts with me in case this weak part of the prototype failed – I just hoped the same problem wouldn't occur in the den! As the time approached for our entry, I went to the toilet to relieve myself before this potentially life changing presentation. But I accidentally 'splashed' myself both whilst pissing and whilst washing my hands. Since I was wearing grey trousers, the dark splashes were very obvious. The crew started to hurry me as I desperately tried to dry off this embarrassment. Imagine being filmed and apparently wetting oneself during the onslaught from the Dragons. This might have made one of the cruellest ever scenes on the Den. I don't know whether my trousers had completely dried by the time Jason and I arrived at the staircase below the den.

The actual staircase is very different to the pretend one shown on TV as one has to spiral up a very tight corkscrew stairwell before emerging into the Den. I went up first, followed by Jason. Climbing the staircase reminded me of my previous struggle against such an apparently trivial part of the show. It was not as bad though, because this time I had a new knee!

Emerging into the Den was quite a surprise, not with the Dragons because they were then already familiar figures, but the set had changed into a green pitch like a football field, much less oppressive than the normal Den. Jason followed me to a predesignated spot with us both standing in front of the Dragons. I was actually quite confident, but looking at Jason I saw a distinct look of dread! Jason and the Dragons exchanged some pleasantries, and the atmosphere calmed down somewhat, but it was obvious that my celebrity supporter was perhaps even more nervous than I was. He was only there for a few minutes at the beginning. There was a distinct sense of awkwardness throughout as though neither the Dragons, Jason nor the production team knew the chemistry or outcome of this celebrity format Dragons' Den. Nevertheless, Jason did a fantastic demonstration of the iiCap which was later shown on TV. The remaining presentation was all down to me. There was a fair bit of banter during the course of the filming, with one clip where Peter Jones borrowed my glasses to try out the invention. Hilariously, we were both blinded – he by the severe distortion caused by my ultra-high index prescription lenses, while everything became a blur for me without my glasses. It was all quite jolly until their decisions when one by one they rejected me, except Duncan Bannatyne who put an extra dose of adrenaline into my heartbeat and a sudden rise of expectation. He started saying something quite optimistic and wishful. Then he said "Kin, I am going to offer you something", Wow, in that few seconds I felt an over-the-moon expectation and exhilaration that I was about to be given an offer of investment. I had asked for £80K for 20% so it was not a big deal given that I had several significant patents. Then, like a sporting referee, he raised his hand to 'offer' me the red card, and simultaneously said "I am out" or something to that extent. There was a roar of laughter from the other Dragons, perhaps the crew members too as this card system was a new special feature of this 'Sporting' edition of Dragons' Den. These were perhaps the biggest contrasting emotions I have ever felt in my life; one moment full of hope and expectation, then suddenly this feeling crushed by a cruel farce. As I left the Den while still on the stairs, I heard another roar of laughter between the Dragons. They may have been just joking between themselves, but at that moment it felt like a double whammy punch to my aspirations. As a BBC assistant came to undo the mic from my jacket, I asked if he had heard Duncan's cruel joke and instinctively said aloud "What a cruel bastard, isn't he?" Later I was interviewed by Evan Davies again. He even remembered my crippled leg from my last appearance and kindly told me to be careful with the stairs!

James, who had waited in another part of the sprawling studio thought I might have got the investment. Never mind, at least I would get some useful publicity when it went on air. But once again, no one could tell me if I would be featured or if it was going to be broadcast at all! The confirmation eventually came only a few weeks before the broadcast in March 2010, which was ten months after the filming! Meanwhile, I had taken the risk to make a minimum order of 1000 iiCaps from Hong Kong. Knowing that my invention would be featured in this prominent show also helped my credibility quite a lot during my quests for investment. I didn't get any direct cash injections, but a couple of local York contacts who barely knew me helped out significantly by building me two websites in preparation for the show. Even more excitingly, my networking led to a joint venture with a Manchester based product design start-up who could potentially develop a promising new accessory related to the iiCap.

With much excitement, I eagerly waited for the broadcast. I was very interested to find out who the other celebrities were as they weren't revealed until the show. They were Greg Rudeski, Iwan Thomas, James Cracknell, Patrick Kielty and Ruby Wax. I would love to have met Ruby Wax, but overall I was lucky to have had Jason Manford who was by far the biggest rising star. I don't think the format was very successful because all the celebrities were rather intimidated by the atmosphere, appearing dumbfounded and not really added much entertaining chemistry or fun to the show. Meanwhile, I was very disappointed that I only got a short clip again. I thought Jason and my encounters with the Dragons were very amusing. Yet again, I missed out on a proper long clip which would have attracted a lot more attention and useful publicity. But most disappointingly, the details of my invention and the investment I sought were not mentioned at all on the Dragons' Den's website, whereas all the other inventions were. I complained to the BBC about this, but the reply was it was all done now so they couldn't amend the website. I later learned from a media related seminar that TV and radio are still exceptionally biased against people who speak with foreign or even strong regional accents. Perhaps that explained my limited exposure. But there might be another reason for the producer to trim down my contribution which I hadn't realised until a month after the show, when Gordon Brown famously had his open mic gaffe against a Rochdale pensioner Gillian Duffy calling her a 'bigoted woman'. My immediate post-filming, instinctive gibe at Duncan Bannatyne being 'a cruel b*****d' may have cost me a more favourable slot.

There was one more twist to the show! There was only one contestant that day who got an investment, the Swiss guy Jean-Claude Baumgartner with the satnav device for skiers. His business was so impressive that there was a battle between the Dragons to win him over. Eventually, the professional businessman got £130K for a 48% joint investment from Peter Jones and Theo Paphitis. Additionally, he even subsequently got the savvy Dragons to loan him a further £100K to patch a momentary cash flow glitch. But Peter and Theo were never to see their money again. As it turned out, Baumgartner's business was in deep trouble and he used the money to fund a luxurious lifestyle instead! As reported by the CPS, "This was a particular brazen and audacious fraud carried out on national television". He was later sentenced for two years and eight months.

When I heard this, I was reminded that during our brief encounter, he probably did meant to give me that latest expensive HTC smart phone in order to rid of the large stock he had already procured. A few days after filming, I got my first iPhone 3GS because I was finally convinced that Apple, with its easy intuitive UI, and amazingly simple to use downloadable 'apps' had completely usurped all other smartphones, while making many separate electronic products redundant. I was surprised that Peter Jones, the mobile phone magnate did not see this coming. If he had, he would not have made that disastrous investment. It was a dark day for the two Dragons. If only… they had given me the money instead!

### Great expectation

2010 was a hopeful year despite a relatively weak direct response from the latest Dragons' Den appearance. A common misconception is to imagine that a single, short TV exposure is going to lead to a fantastic awareness and an upsurge of orders. The reality is that the world is saturated with content from a multitude of sources with the consequence that people are not as easily influenced. One really needs a sustained effort through the various channels to make a difference. Hence, it was necessary to create one's own PR. In that sense I had some luck, with subsequent appearances on several radio programmes, newspapers, some internet forums and winning prizes from several innovation related business competitions. Some orders were coming in through the website, but this free publicity, even high profile PR such as appearing on Dragons' Den only increased web traffic to my sites by a short transient spike. They were no substitute for a proper marketing plan with sustained targeted online and offline advertisements – resources which I didn't have. But there was good news elsewhere, with just £3K worth of government funded innovation vouchers paid to the design department at Sheffield Hallam University, I was able to make a fully working prototype of my injection aid invention. I never thought of this invention as a mass market device, but as a niche product and a stepping stone to a future automated injection aid system. It had great potential to minimise pain and injection failures. Later, the patent which I had submitted a few years earlier was subsequently granted. Even the original smart card alarm business got a boost, my first patent created at the beginning of my invention journey was also granted. In fact, with the help of my distinguished partner and IP expert James, every patent which I had filed passed the rigorous examiners' judgements and was granted patent rights. Altogether, I am now a holder of five patents!

The best news was the 'joint venture' investment I got from a team of Manchester based designers. It had taken me about 6 months of relationship building to finally get the project going. They had seen the usefulness of the iiCap, but they preferred my other accessory idea of creating a universal clip system (which I later called the iiClip) that could be attached to most ordinary caps thus allowing it to function like the iiCap. This could be a mass market accessory to complement the iiCap. I would have preferred if they had wanted to help develop the iiCap, because its 'support system' component part still needed significant professional design and manufacturing input to make it look sexier and for it to be produced reliably. But who was I to dictate to professional designers who thought otherwise? For the first time in a while, my headwear invention ideas had the personal support and positive views of people who were already successful and experienced in the design and sales of sports related products. After going through a series of prototypes over a six month development period, the consumer version of the iiClip finally made it into production. The designers did a great job – and not only in the design; they also provided manufacturing, packaging and marketing assistance. They had even got through to a significant retailer to express intention to place a substantial order of the iiClips. Hence, this provided the confidence and the expectation that the company's first professionally made product[148] would be able to kick start the business. Thus I managed to borrow a significant sum from friends and family to fulfil my part of the bargain – buy the stock, and commit to marketing and selling the product. It was an exciting time, full of optimism and expectation.

But events were starting to unravel along a different path. By the time the stock finally made it onshore in the summer of 2011, the major retailer that had expressed a strong interest had pulled out, while another distributor who had initially showed much support and encouragement changed their mind without good reason. Meanwhile, the buyer door of other major retailers and distributors remained firmly shut despite my persistent attempt to see them. The great expectation had turned into a damp squib. It was very disappointing, not just personally, but also for all the people who had helped me along the way. After invested so much on stock and other business costs, there was just a few thousand pounds left in the bank. I thought briefly about risking it all on one last marketing push, but instinctively put myself and the business in survival mode. A few thousand pounds was not going to go very far in launching a new product. Better to conserve oneself than going bankrupt and losing everything.

## Innovation is the only way

In times like this, I found wise quotes to be quietly comforting.

> *"It does not matter how slowly you go so long as you do not stop."*
>
> ***Confucius***

Of course I would rather go faster, but my inability to raise enough investment made that path impossible. In effect, everything which I had achieved was via bootstrapping from a shoestring budget. But without having obtained a bigger budget, the road to success remained elusive. During this journey, I frequently reflected upon my situation. With the exception of a few years as a post doc, I lived under extraordinary austerity since the beginning of the millennium. But you know what? Although financially poor, and despite many frustrating times during my invention journey, most of the time I was happy and never felt more creative. The excitement of trying to create one's own successful destiny, while inventing new products with potential to benefit many was the driving motivation. My focus was on the journey rather than lamenting on the setback. Furthermore, the flexibility of being one's boss was particularly cherished because of my unstable health condition. The latter had probably made me more unconventional and accepting of financial sacrifice than I might otherwise have been. Perhaps I am easily satisfied; there is a famous saying which I especially appreciated *"Health is the greatest gift. Contentment the greatest wealth"*.

Having said all that, I do still have big ambitions despite failing to achieve my goals. The disappointments, hurdles and falls have undoubted dented my spirit – but somehow, the dents were quickly repaired and I have always managed to push myself further into the challenge again, and again. My company did produce some successes when there was the right person on the team for the job e.g. James for securing IP rights, and similarly Philippe and Heng during the early product development phase. My regret was if only I had a savvy business partner to join me. They could have made a breakthrough sale or gained the company the necessary investments to properly develop and launch the product effectively. I may also have made a common sin of many inventors – pursuing too many ideas at once instead of being totally focused on one invention. I am an entrepreneur, but I lack the single minded focus and salesmanship of a good businessman. Some of my inventions, particularly the iiCap could have been a big success if given a proper chance. Tired and broke, I conceded a temporary break from my invention babies was needed. Hence, I embarked on another route to revive my fortune.

Innovation is the only way. There was one more product which I could invent that wouldn't cost much or rely on other people to make it happen. It is this book. Yes readers, I have created this book purely to promote myself and my inventions. Sorry to have tricked you into buying it.

Seriously, I had wanted to write my story since I was a teenager. Back then I wanted to educate the world about the struggles of living with H. The book could have been a miserable memoir, but fortunately the same condition has also led me to a rather unconventional life and experience – a journey which I hope has entertained and perhaps even enlightened some readers[149].

## 10. INVENT, THEREFORE WE ARE

The great French philosopher Descartes once expounded "I think, therefore I am." It affirms that the very act of thinking and doubting one's existence confirms the existence of the self. Philosophy aside, there is not much doubt that together, all of us through our lives, and through our ancestors contribute to the story of human civilisations. It is a story of the rise and fall of various tribes, states and civilisations, triumphs and tragedies told and untold. This cycle is inevitable, but one common legacy of all of the greatest human achievements is uncovering what were previously the mysterious and unknowns through to the creation of new ideas, knowledge and inventions. Originally, we became inventive because of necessity, literally trying to survive in the hostile wilderness of early Earth. Today, the challenge to survive is relative; in the rich and middle income countries of the world, the major coming struggles are the aging population, housing and the rising cost of healthcare, while for the poorest it is literally surviving in slums or refugee camps fleeing from hunger and wars caused by poor governance, religions, unresolved territorial disputes, cultural differences and over-population. Globally everyone is at risk from catastrophic environmental degradation and major conflicts between countries or ideologies. The challenge ahead for all mankind is as great as ever in human history. Technological innovation can alleviate some of these problems, but it is also inevitable that major social, cultural and personal innovations are required to make the world a better place for all and not just the privileged few. In this final part of the book I hope to offer further thought and suggest solutions to some of humanities ills, focusing on my personal experiences and interests told in this book. Inevitably the exposition is brief but I hope at least some of the ideas are thought provoking.

Many of us would not be alive without advances in medicine and healthcare. I for one would certainly not be here had it not been for modern medicine. But the cost of healthcare is rising inexorably in the developed world while the poorest countries do without even cheap essential drugs. In the UK, healthcare cost to the State is nearing 10% of national GDP, while in the United States, it is 17% and rising[150]. This is expected to rise considerably with the rising costs of the latest treatments and a rapidly aging population. With massive sovereign debts around the world and rising expectations, we need to seriously consider how the sick and vulnerable will be treated. The cost comparison above shows that we must never move into the privatised model to fund healthcare where only the rich and privileged get the best treatments, while the cost to the State is greater. Unlike most commercial businesses where competition is good, I believe the NHS must remain a State utility so it can demand the best prices from pharmaceutical companies, while the cost of patient services are not inflated by the need to generate profits for shareholders. Indeed, global comparisons of healthcare delivery have shown the state funded NHS model is the most cost efficient and provide one of the best healthcare systems in the world. However, soaring healthcare demands mean we cannot be complacent. The NHS must remain dynamic to take advantage of innovative practices or new technology such as tele-medicine which could deliver better and more cost efficient healthcare. Patients themselves need to accept and be educated to be more responsible for their own health. In the same regard, the NHS may need to become less generous in some areas in order to entice greater patient responsibilities and deter wastage such as persistent 'missed appointment' by negligent patients – a topic which I had personally tried to make a difference. On a macro scale, transparency of medicine and service costs must be paramount to minimise inefficient and corrupt practices. Effective regulation is also needed to enable better transparency so that private healthcare providers could not manipulate the system, or distort the market causing greater healthcare costs for all. Globally, countries around the world could work together with pharmaceutical companies to make expensive medicines affordable even in poor countries. While not suggesting countries should enforce state controlled pricing on private companies, as interference of this kind is usually against innovation and counterproductive, we need to think of innovative models of sales and distribution. For example, haemophilia is a rare condition, hence treatment is expensive due to the relatively few patients that need it. Thus, the tragedy is many haemophiliacs around the world remain untreated or have their medication severely rationed. From a personal note, I could not even start to imagine the life of haemophiliacs living in most of the developing world. Many die young, suffer crippling pain and a precarious existence throughout their short lives. If treatment were made available to all the patients in the world, rather than just in rich countries, the unit costs of producing them would be driven down significantly. The cost savings must be passed down to the consumer while big pharma can still maintain a decent profit for further R&Ds. Only global and inter-governmental co-operation could ensure such expanded market. This pricing model is not new – remember how expensive most computer software and games used to be; costing between £20 to £50 per game? Then the App store model changed all that, where the consumers receive the same product for a fraction of what it used to cost. Making essential medicine available for all across the globe should be the prime human right goals for all countries.

The above objective brings forth challenges relating to inequality, incentives and fairness. I am extremely angry and worried at seeing the trend of increasingly excessive polarization of incomes between some company executives, bankers, and even some professionals who are justifying incomes and bonuses that are many multiples of that of the average workers. For example, according to a report in The Telegraph newspaper in 2012, the top 1% of Americans took home 22% of the nation's income and 95% of all income gains since 2009 have gone to the top 1% of the US population. Globally, just 85 top billionaires own as much wealth as the bottom half of the world's population (Oxfam report 2014). I am sick of hearing from some of these people that they are worth it (especially investment bankers who work on the principle of 'privatise the profit, socialise the loss'): company executives earning bonuses while workers and shareholders bear the cut; fund managers creaming off double figure percentages of the value of shares or unit trusts irrespective of the funds' performance; sport stars getting millions while working class fans have to cough up extortionate sum to see them. The same criticism now also applies to some charity organisations and taxpayer funded employees such as some council chiefs, University vice-chancellors, BBC executives and their entertainers demanding ridiculously high salaries and benefits that are based on a cartel of the elite supporting each other's claims to keep their gravy train rolling. Blatantly untrue is these public servants argue their talents are so supposedly unique and rare, so deserving their excessive high reward. Furthermore, despite my huge appreciation to doctors and the medical profession, compared to other advanced economies in Europe I would even go as far as to criticise salaries of senior medics and managers in the NHS for having gone up too far as well, creating an excessive wage burden and differential pay scale which has become the accepted norm. Society in general would be far better served if the average remuneration of senior NHS staffs were more in accordance with other professionals (e.g. senior engineers, senior teachers, University professors) while at the same time ensuring employees are less stressed because the State can afford to employ more of them. The excuse that they need to be paid so much above the average worker to incentivise them is a self-serving and fraudulent claim based on the supposition that people are only motivated by money. The post 2007/2008 austerity has been difficult for many workers, but the relatively low unemployment in this recession has also demonstrated the benefit of the many sharing the pain of austerity: pay stagnation and cuts, instead of mass layoffs and redundancies as experienced in previous recessions. There are of course people who want to be highly rewarded. They could become entrepreneurs, inventors or innovators; people who risk it all, who often work with no salaries for months and even years before success. Even then the reward should come from the success of the products or services which have made real profits, and then rewarded through company shares and dividends only, rather than excessively greedy salaries. Similarly, high ranking executives in successful companies should also be rewarded via a fair share option scheme instead of multi-million pound salaries or bonuses which generally ignore the mid to long term financial performance of the company. If the salaries and rewards of all companies, shareholders and governments were subjected to full transparency and real fairness as described, our world would be a much better place. Hence, the study and resolution of inequality economics and policies must become a priority for all aspects of society, because as much as I love technology and innovations, they are likely to cause major rupture between workers with advanced skills and those without. The rapid advance in IT automation will soon eliminate many more jobs – the most obvious recent example is the rapid adoption of self check out counters in nearly all large retailers. Further advances in Artificial Intelligence and robotic technology will inevitably erode into even middle-class jobs. Governments will eventually have no choice but to impose a wise wage policy which tackles inequality, whilst not harming the creative and entrepreneurial spirit. For example, I would propose a 'maximum take home pay' policy, where employers have total freedom to set the salary of their top employees, but any income above the 'maximum take home pay' is allocated to a special account where the employee must dedicate the sum to socially useful purposes e.g. donate to charities of their choice and/or alternatively invest in start-ups where they could potentially increase their take home income – how much more being dependent on the success of the start-up and on how many new jobs are created.

On the subject of conflicts in human societies, much of it can be reduced to the action of the individual. If every individual applied the principles of respect, tolerance and sensitivity to others, there would be no misery from bullying, exploitation, racism or violence. Of course to some extent the underlying personality traits which lead to these types of behaviour are part of what makes us human. Nevertheless, education can minimise these traits. During my time as a teacher, I was pleasantly surprised to learn that school pupils attend PSHE[151] classes which didn't exist during my school days in the '70s and '80s. In these lessons, they were taught to be more aware of other people's feelings, and to reflect on their own action. As a result, I notice that kids these days, even in the behaviourally challenged schools in which I taught rarely exhibit or tolerate racism and bullying that was so common in my school days. With the exception of war and sectarian torn countries usually caused by nationalism or petty religious divisions, over the five decades which my book spans, the world has overall become a much more tolerant, less violent and liberal place. However, there is a new world in cyber land where prejudice, bullying, sexual perversion, misogyny, racism and hateful indoctrination are common and too often tolerated. One can easily be subjected to such vileness from social media outlets, such as YouTube and Twitter. Even in intelligent media such as in online newspapers, cruel, aggressive and racist comments based on prejudice and ignorance are abundant. One may say this is human nature and free speech, but why should blatantly sexist, racist, homophobic or hateful messages be tolerated? I think this is becoming very corrosive to society and is a step back against progress made in the past three decades. People who post these kinds of messages should lose their anonymity or have their vile messages censored. Transparency and personal responsibilities are needed to enjoy the right to freedom of speech. Much of this problem is an educational and cultural issue, as most hate-related threats, religious bigotry, homophobia or racism which I have witnessed stem from a combination of ignorance, arrogance and readiness to generalise the actions of a few to an entire race or group of people. The best way to counter it is more inclusive education and development of a higher culture, while offensive and aggressive bullies in any media are outed and punished accordingly.

Finally, I'd like to discuss the rise of China which has already altered the world. There have been many positive as well as negative consequences of China's phenomenon growth in the past few decades. In the coming decades, its influence is bound to be even more influential. I belief the positives and negatives will be amplified further depending on China's future strategy and how other countries interact with it. Let me start by listing the negative aspects of China's economic rise. These are massive pollution and environmental degradation, income inequalities, corruption, wild 'east' capitalism which neglects the environment and health and safety to the detriments of millions. The people of China know this very well. The current leaders of China, led by President Xi Jinping have at last responded with a series of tough reforms, in curbing environmental pollution, corruption and stronger rule of law principle. From my analysis, I am more optimistic than pessimistic that China will succeed in tackling the above challenges.

Overall, my opinion is the positives of China's rise far outweigh the negatives listed above. Since Deng Xiaoping's reform which started in 1978, China has lifted more than a billion people out of poverty and under-development. As a result, in the past decade, China's been the main engine of growth for much of the developing countries too, helping and promoting growth in Africa, South America and other poor areas of Asia. The already rich Western countries have also benefited massively through quality and affordable consumer products; increased growth through its exports to China; a faster rate of innovation through increased competition and affordable prototyping/manufacturing; cheap lending rates and massive corporate profits. Unfortunately the

West squandered the last two benefits through its creative innovation of financial products that led to the financial collapse of 2007/8, and in the latter case extreme unfairness to the less skilled workforce by the profits being shared amongst a few instead of increasing prosperity for all.

Globally, the most impressive and under-reported impact of China's rise is her unrivalled contribution to renewable energy investment[152] and infrastructure development, not just within China, but all over Africa, Asia and Latin America. Also, China's demand drove up commodity and energy prices which encouraged China and advanced nations to develop cleaner energy sooner than later, while increasing income to a broad range of developing countries which contributed to further reduction of poverty worldwide. The sudden decline in the price of oil and commodities during 2014/2015 showed the significance of China's prior contribution and importance for future global prosperity as commodity prices must inevitably return to higher sustainable values. The acceleration of development in previously extremely poor regions of the world has naturally reduced high birth rates that would have otherwise led to increasing pressure on the environment. China's environment and that of many of the world's poorer nations stand to benefit by them becoming wealthier since the environment inevitably becomes a higher priority for people and countries with wealth to tackle it. Another area of optimism is as developing countries advance economically; health, education, social progress and the rule of law tend to improve too.

Of course there are further difficult challenges for China and the world. Many in the West resent China's rise, while some countries worry of its influence and power. Part of the animosity is ideological, particularly as China has demonstrated successful development whilst not subscribing to western democratic norms. Meanwhile, some of the biggest concerns are rooted in complex geopolitical and historical legacy issues such as the future of Taiwan, and territorial disputes with India and in the South China Sea. Despite the differences, so long as all parties resolutely continue a pragmatic and ultimately cooperative policy as exemplified by US president Nixon's historic visit to China in 1972, the West, China and other large nations can transform the 21st century into a more prosperous and better place for all. By learning from each other and working together, whilst avoiding escalation of conflicts, we can all look forward to the coming decades where an increasingly developed world population is able to contribute to advancing human civilisation, whilst more inclusive innovations emerge across all disciplines for the benefit of mankind and the environment.

# ENDNOTES

**Chapter 2**

1. Due to its geographical isolation Wenzhou escaped much of the worst effects of these two disasters.
2. The USA also deployed a similar tactic by recruiting Native American speakers in the Second World War.
3. Originally called the Empire Flag, it was built and launched in Newcastle upon Tyne as part of the British Merchant Navy fleet used in the Second World War by the Ministry of War Transport.
4. Of around 1972
5. Probably a member of the 'Emerald beetle' species in English.
6. It could be a 'she' too, but that's extremely rare.
7. In technical jargon, H is therefore an 'X-linked recessive' disorder. Also, a carrier (a female) will tend to have a lower level of F8, which rarely results in any symptom, unless they undergo a major operation.
8. Copyright permission of this diagram was obtained from www.haemnet.com, with thanks to Mike Holland.
9. While it is true that most carriers lead a normal life without suffering from the illness caused by H, it is a popular misconception that they are totally normal as most would require Factor 8 replacement treatment should they find themselves needing an operation or are involved in a serious accident.
10. Actually if my memory is correct, the weapon used was a long wooden ruler.

11. Rebirth is more correct, but my childhood understanding wasn't fully developed to appreciate the difference.
12. It should be noted that effective Factor 8 replacement therapy via Factor 8 concentrates or Factor 8 rich plasma cryo-precipitates only became widely available in the early 70's, and only then in rich countries.
13. With what I know now, a cold bag of peas would have been much more effective than using hot towels as the latter is more likely to inflame the injury further.
14. According to my mum, I may have had one or two doses of Factor 8 earlier in my life. She could not be sure, but as far as I can remember, this was the major event which finally brought my treatment up to date.
15. The HK health service is not as generous as the UK's NHS as there is a small daily charge for hospital stay and for prescription from the family doctor.
16. Wheels on luggage bags were not developed until the 1970's, and the ubiquitous suitcase with a telescopic handle running on two wheels was not invented until 1987, by Robert Plath, a pilot at Northwest airlines. Before then everyone had to endure the hardship of carrying their bags by hand or on their shoulders!
17. When we talk about 'England' in Chinese, it actually means Great Britain or the United Kingdom; this is common colloquial usage, and most non-UK people just don't make the distinction. Similarly, I also use the term Chinese to describe someone with ethnic Chinese identity but not necessarily a person with the nationality of China.

## Chapter 3

18. A rather unreliable Mini as it turned out.
19. In some ways, I am glad that there wasn't the World Wide Web to spoil the surprises of discovering a new place.
20. Of course, my opinions changed as I grew up and understood English and its culture better.

21. Which was a very popular children's TV drama at the time, often depicting disruption and fighting in a fictional London comprehensive school.
22. Not real name, because I can't remember it.
23. Particularly the fight scene in 'Enter the Dragon' where Bruce Lee was knocked down on his back by Sammo Hung, only to immediately bounce back.
24. As I am writing this, I suddenly realised answering this question may actually shed some insight on the different Chinese and English characters. The English kids repeatedly made those expressions not because they were shocked, but to make a point of how they 'ought' to feel. There was a large element of mischievousness in their expressions; most kids were probably just mimicking, and mindlessly joining in the expressive behaviour initiated from the more vocal kids, which in turn led to more laughing and small talk. In other words, English kids, and dare I say, the English in general are less inhibited in expressing their inner thoughts and perception, enjoying group activities and rather liking doing things together. Hence, English culture, despite its fame for individualism has a strong tendency for team spirit and community building. The Chinese kids, on the other hand, behave more seriously and pragmatically. There is nothing out of the ordinary to justify those surprising sounds or expressions. I don't think Chinese kids think any less or more than the English kids, except the former tend to be more reserved and suppressed about expressing themselves. Contrary to popular misconceptions in the West the Chinese is one of the most individualistic people in the world, but the Chinese and Chinese culture in general tend to be more self-centred and consequently have weaker team or community spirit. In fact, I think this is one of the main weaknesses of the Chinese character , and why they are crap at team sport like soccer.
25. This term is used commonly to describe problem joints or muscles that have suffered repeated episodes of bleeds.

26. Cyoprecipitate was a popular and effective substitute for Factor 8 concentrate in the Seventies and early Eighties.
27. Nurse Chin left Manchester shortly after I recovered and we lost touch. Anyone reading this book who knows where she is, please get in touch as I would love to meet her again.
28. The profanity "F**k your mother" is a direct translation of a popular swearing expression in Cantonese.
29. These criticisms are likely to apply to the vast majority of Chinese parents who owned a catering business at the time.
30. Where the upper school was based.
31. Yes, they were just kids as I see them now. But when one is of a similar age, a nasty kid is no less threatening as a nasty adult is to another adult.
32. In the early Eighties, Black protestors devastated part of Liverpool (Toxteth), Manchester (Moss Side) and London (Brixton), due partly to prejudice and racism.
33. The term 'Asian' as used in the UK usually refers to people of South Asian regions e.g. Indians, and Pakistanis, rather than as inferred in the US and most of the rest of the world as people from the continent of Asia. In the US, 'Asian' are most commonly refers to people of East Asia i.e. Chinese, Koreans and Japanese.
34. The doctrine was inspired from Jack London's novel White Fang which had the line 'Respect the strong, pick on the weak' describing the brutality and survival instinct of living in the wilderness.
35. Except for reading about it, especially Willard Price's absolutely wonderful animal adventures.
36. That's "physics" in Chinese.
37. It is often thought that Chinese (or even more generally East Asian) kids in the West tend to pick mathematics and science subjects while eschewing art or humanity subjects. There are several reasons for this; in the case of the first generation immigrant kids, the latter are much harder to pick up or be appreciated in the first place due to the language barrier, so

maths and sciences are more popular because solving maths problems or answering science questions objectively are easier than writing essays. For Western born Chinese kids, the language limitation is not an issue, except their first generation Chinese parents may be more practically minded, which could discourage them to specialise in the arts and humanities.

38. O-levels have since the mid-Eighties been replaced by GCSE's.

39. I know, because one the worst acts of violence I have ever experienced first-hand was from the hand of the deputy head of the school! I was calling out to a friend in the corridor of the school, when suddenly I felt the most almighty smack on the back of my head. Apparently I hadn't noticed the 'keep silence' board for an exam going on in one of the classes. It was a thumping smack of an adult who has applied his full might in the heat of his fury. It almost knocked me off the ground, and as I sat in the classroom, I was feeling the reverberant shock from such a violent attack. I was a stoical kid, but in this particular case I had to concentrate hard to stop the tears flowing because it was so shocking. Naturally, with my H, I was very anxious afterwards. If I could go back in time as an adult, I would definitely hit him back for such a disproportionate act of violence. Fortunately there was no apparent long term effect, probably because I had a thick skull.

40. In the course of writing this book, I have progressively felt less sensitive to the H word. Therefore, depending on my mood for the rest of the book there will be less consistency about avoiding the whole word!

41. There were other theories that UFOs were in fact time-machines from the future. At 15, I wrote an amateurish 'scientific' article about this, submitted to UFO review proposing this idea, which explains why UFO activities tend to be discreet so as to minimise altering history, and

observations of most aliens are humanoids, even sometimes with 'Oriental features'. At that time, I thought there was hope that they might be future Chinese, once again being at the forefront of science and technology!

42. I remember being really annoyed when in an undergraduate physics lecture on magnetism, all reference to the compass being used was credited to the Greeks.

43. The great English philosopher and scientist Francis Bacon had selected three inventions; paper and printing, gunpowder, and the magnetic compass which he considered as having the greatest influence in transforming Europe's Middle Ages to the modern era. He regarded these inventions as 'obscure and inglorious', without ever realising all of them were of Chinese origin. There are similarly many other great and influential Chinese inventions and ideas over the antiquity and Middle ages of which most people are unaware. For those who are interested in finding out more, I recommend the work of Robert K.G. Temple's book "China: Land of discovery and invention".

44. Surely the best Western food is Italian, with Sichuan cuisine the best in the world!

45. As opposed to the Eurocentric curriculum which I was later subjected to, where almost all significant historical figures taught were White Europeans or Americans.

46. My first language was actually the Wenzhou dialect, my parents' mother tongue, but growing up in HK meant Cantonese became my most fluent language.

47. I have deliberately been careful to avoid the cliché of talking about 'the different thinking between the West and China', because that would have implied a lack of diversity of opinions in both Westerners and the Chinese people.

## Chapter 4

48. As with most of Einstein's discoveries, the conjecture of light quanta revolutionised understanding of light not just

as a wave, but also as a packet of energy like a particle. This discovery made Einstein one of the fathers of quantum theory, along with Max Planck, and Neils Bohr.

49. With the discovery of the Special theory of Relativity, space and time could no longer be perceived as separate entities, but rather as a union of spacetime. Ironically, Einstein did not realise this significance immediately after his discovery; it was Einstein's former teacher Hermann Minkowski in 1908 who formally developed the concept of spacetime, which was crucially important in the subsequent development of the so called theory of General Relativity.

50. I think it was Jeremy Bernstein's biography of Einstein.

51. Einstein, being a theoretical physicist of course never did any such experiment. He thought of many of his greatest ideas by doing 'gedankenexperiments' or thought experiment.

52. I have based my observation on the Upper school only which catered for students from Year 10 to Year 13 (Years 12 and 13 being the sixth form age group). Younger year groups were based in another site half a mile away.

53. I heard rumours that he would get his hot meal lunch in the dining hall, and then take it to his office to consume it, rather than eating it in the dining hall with the other teachers.

54. I think it was a physics lesson.

55. The term 'devil' (or 'ghost' as translated literally) is a derogatory noun in common use in Southern China when referring to white people and other foreigners. Its usage is in many ways similar to non-Chinese calling Chinese people, 'Chinaman'; in the sense that it is often not meant to be offensive, other than being lazy and ignorant of its negative connotation. Of course, I abhor both expressions in today's world.

56. Back in the 80's, I had actually really wanted to know some universal racist slurs for white people, so I might use them against the white racists I frequently encountered then, and to see their reaction. But, I just couldn't discover any, because

one just doesn't hear them. Even today, with the help of the internet (e.g. via Wikipedia), the supposed racist slurs such as white trash, whitey etc. never really have the same vitriolic impact as those against non-white people. The reason is not because white people are more tolerant and care less about being offended by 'mere' words. The reason is 'white privilege', a term which I have only learned about in the course of writing this book. I urge everyone to look up this term to fully comprehend what racism is about.

57. Of course, this is increasingly less true now for fast developing China and India.

58. Paraphrasing President Ronald Reagan.

59. Saudi Arabia is the main sponsor of Wahhabi Islam, promoted around the Sunni Muslim world. Note, my earlier writing of my school friend Ikhlaq was purely based on my childhood perception, without any pretence of deep understanding of his background. Hence it is quite possible with hindsight that his fondness for wishing he was a Saudi could have been influenced by Saudi funded propaganda which was probably already prevalent in Oldham's mosques in the 80's.

60. Wars and conflicts are of course by no means unique to Islam. In the '80's and '90's I recall getting particularly tiresome with daily news of sectarian conflicts between Catholics and Protestants in Northern Ireland.

61. I wish to emphasis to my Pakistani and Muslim friends that this section, like the rest of this book is a frank exposition of my experience and analysis. Sensitive points have been made, and some might be offensive to some. I hope the only people who are offended are bigots and religious fascists, not my liberal Pakistanis and Muslim friends who I value and respect.

62. This is entirely my perception and is probably without any scientific merit!

63. "Murder in the playground: The Burnage report" (1990)is a report by Ian MacDonald QC into racism and racial violence in Manchester schools.

## Chapter 5

64. You may recall that I felt extremely lucky to have gone to Burnage high school for my A-levels, which despite having its troubles somehow produced one of the best exam results in Manchester!
65. 1.5G means the one and a half generation immigrants who came to their new country before they became a teenager.
66. The one clear racist incident that I still remember quite vividly was in the very last week of my undergraduate University course – a drunken student amongst his friends walking past the ground floor of the physics computer room shouted racist slurs at me for no apparent reason. After almost 3 years of non-racist encounters, it was a big shock to hear them again. My immediate reaction was to ignore him, but as he disappeared past, I became very angry that I didn't respond more aggressively to such abuse. This incident made me realised just how pleasant living without casual racism really was and highlighted to me just how much I deeply regretted my previous existence of tolerating almost daily racist abuse.
67. By the late Eighties, it was realised that non-A and non-B hepatitis were in fact caused by something which was far more serious than initially thought.
68. There is a perception that students are constantly having great sex with new partners. The reality is that a substantial percentage of my English and non-English friends at the time had a rather mundane sexless lives! Admittedly, many of my English friends were geeky physicists.
69. Via his famous 1905 paper on the photoelectric effect where he postulated the existence of light as a particle object (i.e. photons).
70. A reasonable and precise definition of physical reality was proposed in the EPR paper; "If, without in any way disturbing a system, we can predict with certainty the value of a physical quantity, then there exists an element of physical reality corresponding to this physical quantity".

71. Realism is the philosophical doctrine that the world is made up of objects whose existence is independent of human experience. Realism is therefore the criterion in EPR's definition of a 'physical reality'.
72. Don't worry if you don't totally understand this section – it took me many hours of re-reading and re-reflecting on the EPR problems before I understood it again! The brilliant American physicist Richard Feynman once said" I think it is safe to say that no one understands quantum mechanics".
73. Greenberger-Horne-Zeilinger
74. The numbness was caused by the compression of the nerve by the compressed muscles. Later I learned this kind of injury could have led to me losing the function of my arm.
75. I must have had a treatment back in Manchester in order to recover from such a heavy bleed.
76. Zhou lost his position after June 4th, and was subsequently under house arrest for the rest of his life.
77. As shown in 'The Gate of Heavenly Peace', a 1995 documentary about the event of Tiananmen Square in 1989.
78. One of these interviews was with Prof. Peter Higgs (of Higgs' particle fame and 2013 physics Nobel prize winner) in Edinburgh University.
79. A few years later, funding opportunities in this field greatly expanded because of its link with quantum computing.
80. Apologies to Dr Hillel of Manchester University who found the funding. He must have been disappointed by my decision.
81. Actually there was another targeted joint problem with my left elbow, but I won't bother the reader with every one of my troubles!
82. One anecdote I remember is that Josephson used to carry with him the Nobel prize medal all the time, so he could show it to people who enquired about it.
83. I learnt from experience to give Prof Hawking plenty of room in the narrow corridors of the department as he always seemed to drive his wheelchair at top speed; at one time,

his heavy electric wheelchair missing my foot by just a few millimetres. In the popular press, Hawking is often compared with the likes of Einstein and Newton, but this is really a pop culture exaggeration. While Hawking is brilliant and certainly deserving of his achievements, especially in view of his profound disability, if I had to rank Hawking, he would reside in the 1st division of scintillating minds, while people like Newton, Einstein and Gauss would be in a super premier league, whereas Maxwell and Dirac would be in the premier league.

84. It was purported in the news media that the East German leadership had considered the so called 'Tiananmen Square' solution to regain control. Wisely, they didn't. I like to think the tragic event in China five months earlier, and its subsequent estrangement had deterred the East German authority from taking the violent option.

85. Exasperated blood in injured joints is extremely corrosive to cartilage and bones. Hence many severe H's suffer from severe premature arthritis as a result of repeated bleeds in their joints.

86. I think it was around the time of this realisation that I developed a desire to tell my story – to warn other young H's to look after themselves properly and to learn from my mistakes.

87. The German word Schadenfreude is one of my favourite words because I have frequently reflected on its meaning to make myself feel how lucky I am when compared to the desperate suffering of others.

## Chapter 6

88. Which is part of the United Kingdom Atomic Energy Authority (or UKAEA).

89. I am not saying that employers might not have offered the job because of discrimination (which is illegal), but in a very competitive situation where two candidates, one with normal

health status, and one with a medical condition are of almost of equal worthiness and likeability; whom would you choose?

90. I still recall this dream quite vividly – that of an old man (probably Einstein) sitting on a chair contemplating problems in his head, and me in the dream wishing one day that I would be a 'nuclear' physicist like him. Note the discipline of nuclear physics is very broad. It is usually referred to as the study of the nucleus of the atom – which is actually a branch of physics which I found to be most boring and uninspiring. Einstein was intimately related with nuclear physics through his famous $E=mc^2$ equation, which he had deduced through analysing the nature of light, space and time rather than the study of the nucleus itself.

91. The speed of light is approximately 300,000 kilometres per second.

92. For readers who are more technically minded, the reaction is $D^2 + T^3 \rightarrow He^4$ (3.5MeV)+$n^1$(14.1 MeV)

93. Note, if the plasma for whatever reason does escape the confinement of the magnetic fields and hits the wall, the outcome is a rapid cooling down of the D-T plasma fuel with the immediate effect of safely shutting down the fusion reaction.

94. Bryan Taylor is also known for being the co-developer of Britain's first hydrogen bomb. I remember an interesting anecdote by Bryan one lunch time recounting how he used to play with a plutonium ball with his bare hands during his time at the Atomic Weapons Establishment at Aldermaston.

95. ERM is the European Exchange Rate Mechanism.

96. The likely reason was pride and ignorance as at the time, the thought of owning crutches to help me out had simply not cross my mind, although with hindsight, I believe the doctors should have advised me of this possibility much earlier.

97. Whenever possible, I tried to mitigate my sick absence by doing or thinking about work. Fortunately this was possible in my job as a theoretician, although the lack of internet and World Wide Web at the time prevented effective work.

98. Note, due to strict copyright law, I could not write out the lyrics in its full original meaningful sentence. I urge the reader to seek out the lyrics of these songs (e.g. from web resources) and fill in the missing words to get the inspiring feelings I felt.
99. It was of course the famous One Love Peace Concert on the eve of the Jamaican election in April 1978 where the two men were the opposing politicians Edward Seaga of the Jamaica Labour Party (supported by the CIA) and Michael Manley of the People's National Party (PNP) supported by Cuba.
100. These are common acupressure balls often found in traditional Chinese products shops. At the time I was trying all kinds of different products to improve my health situation.
101. Perhaps this is the state of Nirvana as used in Buddhist terminology.
102. I think he had a black belt in Judo obtained during his University days.
103. If the GDP is measured taking into account 'Purchasing Power Parity', China became the world's largest economy in 2014.
104. This was despite the last governor of HK, Chris Patten who pushed forward new democratic reforms without consultation with China. It was an extremely provocative act that was supposed to be for the good of the HK people. It led to a bitter breakdown of communication between China and the departing colonial administration, and served to worsen relations with Britain. Furthermore, Hong Kong people became more polarised. As an onlooker who cares about HK, I found Patten's 11th hour reforms sanctimonious and dangerous, so typical of politicians who take to grandstanding in the name of democracy, instead of fully considering the consequences.

## Chapter 7

105. The funding situation has changed completely now since the discovery of the application to quantum computing which is closely linked to the subject.

106. Discussed in chapter 5.
107. Actually, my average annual sick leave during the time at Culham was probably just between 5 to 10 days a year, having often covered up H-related illness by using up some of my holiday leave instead.
108. Learner riders were simply taught to stay in the middle of our part of the road. The more proactive style of riding was probably not taught for beginners because it might cause more confusion.
109. Note, one should always ride within his or her limit, as I have always done. Being pressured to go beyond the limit is an invitation to sample hospital food or worse, be with worms 6 foot under.
110. More of that later.
111. For privacy reasons, Rob and Suzy are not their real names.
112. Coincidentally, while writing this part of this book, I was in the middle of reading a very harrowing book called "April Fool's Day" by Australia's most famous author Bryce Courtenay. It is a real story about his haemophiliac son Damon who died from AIDs on the same day, a decade earlier.
113. He is currently serving a 20 year jail term for corruption during his presidency between 2000 to 2008.
114. With deep sadness, subsequent leading western leaders have simply not learned anything about the insanity of military and political interventions in other peoples' civil wars. Western imprudent supports against their so called enemies have emphatically magnified tragedies in Libya, Syria and Ukraine. The rise of the fascist Islamic State (or ISIL) in the Middle East is another terrible legacy of biased and severely misguided foreign policies.
115. With respect to the people of Scotland and Wales, for most foreigners not familiar with British history or geography, English and British, or England and Britain are often thought as synonymous.
116. Depending where one is, indigenous Britons mean White English, Scottish or Welsh people.

117. In the US, it is increasingly common to refer to young immigrants who arrived before their teenage years as the one and a half generation (or 1.5G). I like this term because our experience and development mindset is quite distinct from the older first generation migrants or those of the second generation.
118. By 'openness', I don't particular mean openness to immigration, but rather the trait of being open to new cultures and ideas while also being prolific in sharing it's exceptional cultural and scientific achievements.
119. It is outrageous that in the 21st century one still does see frequent 'casual' racial slurs in Western media. A particular personal example is the word "Chinaman" which is still frequently used in the British media. Worst culprits were BBC sport commentators when referring to Chinese athletes, and in print, seeing "Johnny Chinaman" repeatedly made by Jeremy Clarkson in his Sunday Times newspaper column as though it was perfectly proper, when Chinese or Chinese man would be entirely normal to use. There are some people who don't see a problem because they are ignorant of the negative connotations, and I am happy to correct them. However, for highly paid and professional journalists to continue perpetuating this pejorative term is disgraceful.

## Chapter 8

120. "Did Not Attends" is a common jargon used by health services for missed appointments.
121. The invention idea was an electrical powered cooling pad for the purpose of reducing swellings after a bleed. The idea was enthusiastically taken up by an invention promotion company, American Inventors Corp or AIC. The company's service was useless and costs almost £2000 of my parents' savings. Much later in 1999, I discovered that 10 people from AIC was indicted in a $60 million invention promotion scam. After this painful lesson, I focused on physics and didn't think about inventing until almost two decades later!

122. For the younger readers: in those days, the only way to find out is through the arduous journey of visiting a commercial library to find the relevant resource or contact details. Write to or call the relevant companies or experts, who may then give you just enough information to move on. If you needed more information, you probably had to request a brochure or pay for the product.
123. A patent can only be granted for an invention which has not been previously divulged in the public domain. Discussion of the invention under a clearly defined confidential agreement is fine, but there is no guarantee that the invention idea may not leak out.
124. Recall that Maxwell equations completely describe the propagation of an electromagnetic wave which consists of an electric and magnetic field components.
125. It was programming language C to be precise, and I just can't get my head round those pointers stuff. You will get this if you have ever programmed in C.
126. I had been feeling regular 'arthritic' type pains for a while by then. With hindsight, I now know they were most likely caused by bleeds instead of arthritis. This is far more serious than standard arthritis because the destructive effect of bleeds in the joints is much faster.
127. In 2000, James Dyson successfully sued Hoover in a patent litigation dispute because there was one crucial paragraph of technical improvement in his patent, without which he would have lost the case, and probably lose his business too.
128. Sadly, Businesslink and Yorkshire Forward no longer exist.
129. There are of course a group of innovators who probably contribute most to the world of innovation; they are the scientists, researchers and engineers working in academia, research laboratories or large companies. Indeed many spinoff companies have originated from such environments, with some combining the best of both worlds; the world of a secure paid position with the excitement of additional entrepreneurial opportunities. Unfortunately, this was not an

option for me as none of my inventions were related to my previous careers.
130. I get really angry when I hear this. I almost wanted to tell one investor to piss off and stop pretending to be an investor.
131. Unfortunately, only for those who were alive.
132. Since then, I have bought a pair of compact crutches which provide a much stronger and even weighted support.
133. Source: "Jack The Great Seducer" by Edward Douglas. Publisher: HarperCollins.
134. For more of the story, see March 17, 1999|Cyndia Zwahlen, Special to the LA Times, "Helping Cardholders Read the Fine Print".
135. Actually, when I read about the Lenscard story in 2006, my own card project was already entering the 4th year.
136. As a suggestion to avoiding future mishaps like this, I propose that all future similar situations be doubly verified before a definitive answer is given out.
137. Rather irrelevantly, but ironically, the mouse protein was originally derived from Chinese hamster cells!
138. In the eighties, Wang Computer was a global multi-billion dollar computer company which even had a large tower building presence in the centre of Manchester.
139. Source: Gibson Index Ltd www.gibson-index.com Imperial College, Manchester and a few Scottish Universities have a clutch of high quality spinouts started by ethnic Chinese Scientists; based on Gibson's independent assessment of the top 30 spinout companies in the UK.
140. The speaker was Matthew Driver, owner and MD of NetworkChina.
141. The Archer Inquiry is an independent public inquiry report on NHS supplied contaminated blood and blood products; published 23 February 2009.
142. Most H including myself were on human derived F8 until 2004. These F8 were heat treated to deactivate viruses such as HIV and HCV, but one could not be 100% confident until synthetic F8 was made available.

143. Tragically, the cancer returned later in the year. She died at home in 2011. She was just 60 years old.
144. During September to December of 2014, HK experienced dramatic political protests from mostly young people demanding universal suffrage for the selection (i.e. public nomination) of HK's Chief Executive. There is clearly a significant disconnect of some HK people, particularly the younger populace with China's influence. While these demonstrators need to be heard, the majority of HK people and my view is China has fundamentally abide with the Basic Law which enshrined the 'One country, two systems' principle. The exemplary and peaceful manners in which the protests ended have assured me of the competency of both HK's and China's governments in resolving future political protests and demands.

**Chapter 9**

145. I might be totally mistaken about this!
146. Perhaps incited by the memory of a particular BBC report of an Iraqi haemophiliac boy injured and debilitated in a barely functioning Bagdad hospital in the second Iraq war a few years earlier.
147. It took me nearly four months after our relationship started before I could reveal my H-status.
148. Although with the benefit of hindsight, I regret some design decisions which made the iiClip not as user friendly, or as generally effective as it could have been.
149. During the course of writing this book, I became a part-time teacher to supplement my income. It was an amazing experience with quite a few woes and many entertaining anecdotes. I hope to expand on it another time.

**Chapter 10**

150. 2014 figures.
151. Which stands for Personal, Social, and Health Education.
152. RE100 China Analysis report, April 2015. Download available from www.theclimategroup.org

# Acknowledgements

I wish to thank writers Anna Chen, Paul Mills and Tara Moore for their generous feedback and professional advice. I am also particularly indebted to Dr June Hargreaves MBE and Dr Stephen Oxley for their critical review of the draft.

## About the Author

**Dr Kin F. Kam** abandoned a conventional science career to become an independent inventor and writer. He subsequently obtained five patents for several consumer and medical related innovations, some of which have been commercialised. A graduate of Cambridge and York Universities, he was formerly a research physicist having worked in quantum theory, nuclear fusion and computational electromagnetics.

Forever curious about nature and current affairs, Kin welcomes active engagement on Twitter (@drkinkam). An animal lover, he hopes to keep more pets in the not too distant future.

Please feel free to ask Kin questions or discuss this book via the twitter hashtag #BloodDragonsAndLions

www.drkinkam.com

#BloodDragonsAndLions